This report contains the collective views of an international group of experts and does not necessarily represent the decisions or the stated policy of the United Nations Environment Programme, the International Labour Organisation, or the World Health Organization

Environmental Health Criteria 84

2,4-DICHLOROPHENOXYACETIC ACID (2,4-D) – ENVIRONMENTAL ASPECTS

Published under the joint sponsorship of
the United Nations Environment Programme,
the International Labour Organisation,
and the World Health Organization

World Health Organization
Geneva, 1989

The **International Programme on Chemical Safety (IPCS)** is a joint venture of the United Nations Environment Programme, the International Labour Organisation, and the World Health Organization. The main objective of the IPCS is to carry out and disseminate evaluations of the effects of chemicals on human health and the quality of the environment. Supporting activities include the development of epidemiological, experimental laboratory, and risk-assessment methods that could produce internationally comparable results, and the development of manpower in the field of toxicology. Other activities carried out by the IPCS include the development of know-how for coping with chemical accidents, coordination of laboratory testing and epidemiological studies, and promotion of research on the mechanisms of the biological action of chemicals.

ISBN 92 4 154284 5

ISSN 0250-863X

PRINTED IN FINLAND

88/7853 — VAMMALA — 5000

CONTENTS

ENVIRONMENTAL HEALTH CRITERIA FOR
2,4-DICHLOROPHENOXYACETIC ACID (2,4-D) -
ENVIRONMENTAL ASPECTS

WHO TASK GROUP ON ENVIRONMENTAL HEALTH CRITERIA FOR 2,4-DICHLOROPHENOXYACETIC ACID (2,4-D) - ENVIRONMENTAL ASPECTS

Members

Dr. L.A. Albert, Director, Environmental Pollution Programme, National Institute for Research on Biotic Resources, Veracruz, Mexico

Mr. H. Craven, Ecological Effects Branch, Office of Pesticides Programs, US Environmental Protection Agency, Washington DC, USA

Dr. A.H. El-Sebae, Division of Pesticide Toxicology, Faculty of Agriculture, Alexandria University, Alexandria, Egypt

Dr. J.W. Everts, Department of Toxicology, Agricultural University, Wageningen, Netherlands

Dr. W. Fabig, Fraunhofer Institute for Environmental Chemistry and Ecotoxicology, Schmallenberg-Grafschaft, Federal Republic of Germany

Dr. R. Koch, Division of Toxicology, Research Institute for Hygiene and Microbiology, Bad Elster, German Democratic Republic *(Chairman)*

Dr. Y. Kurokawa, Division of Toxicology, Biological Safety Research Centre, National Institute of Hygienic Sciences, Tokyo, Japan

Dr. E.D. Magallona, Pesticide Toxicology and Chemistry Laboratory, University of the Philippines at Los Banos, College of Agriculture, Laguna, Philippines

Professor P.N. Viswanathan, Ecotoxicology Section, Industrial Toxicology Research Centre, Lucknow, India

Observers

Dr. M.A.S. Burton, The Monitoring and Assessment Research Centre, London, United Kingdom

Dr. I. Newton, The Institute of Terrestrial Ecology, Monks Wood Experimental Station, Huntingdon, United Kingdom

Secretariat

Dr. S. Dobson, The Institute of Terrestrial Ecology, Monks Wood Experimental Station, Huntingdon, United Kingdom *(Rapporteur)*

Dr. M. Gilbert, International Programme on Chemical Safety, World Health Organization, Geneva, Switzerland *(Secretary)*

Mr. P.D. Howe, The Institute of Terrestrial Ecology, Monks Wood Experimental Station, Huntingdon, United Kingdom

NOTE TO READERS OF THE CRITERIA DOCUMENTS

Every effort has been made to present information in the criteria documents as accurately as possible without unduly delaying their publication. In the interest of all users of the environmental health criteria documents, readers are kindly requested to communicate any errors that may have occurred to the Manager of the International Programme on Chemical Safety, World Health Organization, Geneva, Switzerland, in order that they may be included in corrigenda, which will appear in subsequent volumes.

* * *

A detailed data profile and a legal file can be obtained from the International Register of Potentially Toxic Chemicals, Palais des Nations, 1211 Geneva 10, Switzerland (Telephone no. 988400 - 985850).

ENVIRONMENTAL HEALTH CRITERIA FOR 2,4-DICHLOROPHENOXYACETIC ACID (2,4-D) - ENVIRONMENTAL ASPECTS

A WHO Task Group on Environmental Health Criteria for 2,4-Dichlorophenoxyacetic acid (2,4-D) - Environmental Aspects met at the Institute of Terrestrial Ecology, Monks Wood, United Kingdom, from 14 to 18 December 1987. Dr I. Newton welcomed the participants on behalf of the host institution, and Dr M. Gilbert opened the meeting on behalf of the three co-sponsoring organizations of the IPCS (ILO/UNEP/WHO). The Task Group reviewed and revised the draft criteria document and made an evaluation of the risks for the environment from exposure to 2,4-D.

The first draft of this document was prepared by Dr S. Dobson and Mr P.D. Howe, Institute of Terrestrial Ecology. Dr M. Gilbert and Dr P.G. Jenkins, both members of the IPCS Central Unit, were responsible for the overall scientific content and editing, respectively.

* * *

Partial financial support for the publication of this criteria document was kindly provided by the United States Department of Health and Human Services, through a contract from the National Institute of Environmental Health Sciences, Research Triangle Park, North Carolina, USA - a WHO Collaborating Centre for Environmental Health Effects.

INTRODUCTION

There is a fundamental difference in approach between the toxicologist and the ecotoxicologist concerning the appraisal of the potential threat posed by chemicals. The toxicologist, because his concern is with human health and welfare, is preoccupied with any adverse effects on individuals, whether or not they have ultimate effects on performance or survival. The ecotoxicologist, in contrast, is concerned primarily with the maintenance of population levels of organisms in the environment. In toxicity tests, he is interested in effects on the performance of individuals - in their reproduction and survival - only insofar as these might ultimately affect the population size. To him, minor biochemical and physiological effects of toxicants are irrelevant if they do not, in turn, affect reproduction, growth, or survival.

It is the aim of this document to take the ecotoxicologist's point of view and consider effects on populations of organisms in the environment. The risk to human health of the use of 2,4-D was evaluated in Environmental Health Criteria 29: 2,4-Dichlorophenoxyacetic acid (WHO, 1984). This document did not consider effects on organisms in the environment, but did consider environmental levels of 2,4-D likely to arise from recommended uses. No attempt has been made here to reassess the human health risk; the interested reader should refer to the original document, which contains the relevant literature in this area.

This document, although based on a thorough survey of the literature, is not intended to be exhaustive in the material included. In order to keep the document concise, only those data which were considered to be essential in the evaluation of the risk posed by 2,4-D to the environment have been included.

The term bioaccumulation indicates that organisms take up chemicals to a greater concentration than that found in their environment or their food. 'Bioconcentration factor' is a quantitative way of expressing bioaccumulation: the ratio of the concentration of the chemical in the organism to the concentration of the chemical in the environment or food. Biomagnification refers, in this document, to the progressive accumulation of chemicals along a food chain.

1. SUMMARY AND CONCLUSIONS

2,4-D is a selective herbicide which kills broad-leaved plants but not grasses or conifers. Its chemical structure is a modification of a naturally occuring plant hormone. 2,4-D is available as the free acid but is used, in agriculture and forestry, in formulations as a salt or ester.

1.1 Uptake, Accumulation, Elimination, and Biodegradation

2,4-D does not persist in soil because of its rapid degradation.

The physico-chemical properties of 2,4-D acid and its formulations have an important effect on its behaviour in environmental compartments.

The bioavailability to, and uptake by, aquatic and terrestrial organisms is strongly influenced by the organic matter content of soils, microbiological activity, and by environmental conditions such as temperature and pH. Although highly inconsistent, the data on dissipation and bioavailability in various soils demonstrate a marked influence of differences in the texture and mineral composition of the soil. In aerobic soils, with a high content of organic material, and at high pH values and temperatures, toxic effects are limited because of rapid degradation of 2,4-D.

Uptake is followed by rapid excretion in most organisms. With the exception of some algae, the retention of 2,4-D by organisms in the environment cannot be expected, because of its rapid degradation.

Some microorganisms are capable of utilizing 2,4-D as their sole carbon source. Repeated application to soil stimulates the number of organisms capable of degrading the compound.

1.2 Toxicity to Microorganisms

In general 2,4-D is relatively non-toxic to water and soil microorganisms at recommended field application rates. No effect of 2,4-D was recorded on 17 genera of freshwater and two genera of marine algae at concentrations up to 222 mg/litre. No effect of 2,4-D on respiration of either sandy loam or clay loam soils was observed at concentrations up to 200 mg/kg.

N-fixation by aquatic algae is affected at high concentrations of 2,4-D acid (400 mg/litre). An effect of 2,4-D esters on N-fixation occurs from a concentration of 36 mg/litre upwards. N-fixing algae in topsoils appear to be more vulnerable to 2,4-D acid than other algal species. The Cyanobacteria (blue-green algae) are important as the major N_2 source in tropical ponds and soils.

In the range of 25.2 to 50.4 mg/litre, 2,4-D was inhibitory to all types of soil fungi.

Cell division was reduced in a green alga by 2,4-D at 20 mg/litre and stopped at 50 mg/litre. No effect was observed on a natural phytoplankton community after exposure to 2,4-D at 1 mg/litre. However, exposure to esters of 2,4-D reduced productivity in these organisms.

1.3 Toxicity to Aquatic Organisms

At recommended application rates, the concentration of 2,4-D in water has been estimated to be a maximum of 50 mg/litre. Most applications would lead to water concentrations much lower than this (between 0.1 and 1.0 mg/litre).

The short-term toxicity data on the effects of 2,4-D free acid, its salts, and esters on aquatic invertebrates is extensive. Ester formulations are more toxic than the free acids or salts. Sensitivity variations exist among species in response to the same formulation. Organisms become more sensitive to 2,4-D when the water temperature increases. Reproductive impairment occurred at concentrations below 0.1 of the short-term toxic levels determined for these formulations.

LC_{50} values for fish vary considerably. This variation is partly due to differences in species sensitivity, chemical structure (esters, salts, or free acid), and formulation of the herbicide.

Although the free acid is the physiologically toxic entity, the ester formulations represent a major hazard to fish when used directly as aquatic herbicides (because they are more readily taken up by fish). Amine salt formulations used to control aquatic weeds do not affect adult fish.

The no-observed-effect-level (NOEL) varies with the species and the formulation: less than 1 mg/litre (coho salmon) to 50 mg/litre (rainbow trout).

Fish larvae are the most sensitive life stage but are unlikely to be affected under normal usage of the herbicide.

Long-term adverse effects on fish are observed only at concentrations higher than those produced after 2,4-D has been applied at recommended rates.

Few studies are related to the effects of environmental variables, such as temperature and water hardness, on 2,4-D toxicity to fish. Higher temperature possibly increases the toxicity. This might be considered when assessing the safety of 2,4-D to fish during control of aquatic weeds.

Fish detect and avoid 2,4-D only at higher concentrations than those obtained under normal conditions of use.

Amphibian larvae are generally tolerant to amine salts of 2,4-D; the 96-h LC_{50} values exceed 100 mg/litre. Of the species tested, only one was sensitive. No information is available on reproductive development and differentiation or on tissue levels.

1.4 Toxicity to Terrestrial Organisms

Based on the widespread use of 2,4-D and its formulations, insects of many kinds could be exposed to the material. Although the compounds are generally classified as non-toxic for beneficial insects, such as honey bees and natural enemies of pests, some adverse effects have been reported on the early life-stages and adults of some insects.

Esters are less toxic to insects than are salts or the free acid.

Birds, and particularly the eggs of ground-nesting species, would be exposed to 2,4-D after spraying. Food items could also be expected to be contaminated by the herbicide. However, most studies on birds and their eggs have been conducted at exposures far higher than could be expected in the field.

LD_{50} values from acute oral and from short-term dietary dosing indicate low toxicity of 2,4-D to birds. In longer-term studies, effects have only been reported at extremely high exposures (for example, kidney effects after dosing in drinking water with concentrations in excess of the solubility of the material). There have been no reported effects on reproductive parameters, even at excessive exposure levels.

A single study reported adverse effects on the embryos of birds' eggs sprayed with 2,4-D. Many studies since have shown no effect on hatchability of eggs and no increased incidence of abnormalities in chicks even after very high exposure to 2,4-D. Other work indicates a very poor penetration of the eggshell by the herbicide. It can only be concluded that after normal, or even after excessive, 2,4-D use, there would be no effect on birds' eggs.

Based on the available data, no generalization can be made about the hazard of 2,4-D to mammals in the field. Data on voles indicate that the herbicide poses no hazard.

1.5 Effects of 2,4-D in the Field

No direct toxic effects, acute or long-term, of 2,4-D applications under field conditions on any animals species have been observed thus far.

There are, inevitably, indirect effects resulting from the intended selective herbicidal properties of the compound. These effects would result from the use of any herbicide or from other methods of land management. There will, therefore, be effects for mammals, birds, and insects because of food deprivation, modification of habitat, requirements for nesting, shelter, etc.

The application of 2,4-D appears to be less hazardous to the beneficial epigeal arthropod community than physical cultivation.

2. PHYSICAL AND CHEMICAL PROPERTIES

Details of the physical and chemical properties of 2,4-dichloropheoxyacetic acid (2,4-D) are given in Environmental Health Criteria 29: 2,4-D (WHO, 1984). The relevant chapter is summarized here.

The structures of 2,4-D and of chemically-related phenoxy herbicides in common use are given in Fig. 1. 2,4-D is a chlorinated form of a natural plant hormone (auxin).

Some physical properties of 2,4-D and of the 2,4-D derivatives that are used in agriculture are summarized in Tables 1 & 2.

2,4-D has growth-regulating and herbicidal properties in broad-leaved plants. Because of its solubility, 2,4-D is rarely used in the form of the acid; commercial 2,4-D herbicide formulations consist of the more soluble forms such as alkali salts, amine salts, or esters. These are combined with solvents, carriers, or surfactants and are marketed in the form of dusts, granules, emulsions, or oil and water solutions in a wide range of concentrations.

Table 1. Physical properties of 2,4-D

Molecular formula	$C_8H_6Cl_2O_3$
Relative molecular mass	221.0
Melting point	140 - 141 °C
Solubility in water	slightly soluble
Solubility in organic solvents	soluble
Vapour pressure	52.3 Pa at 160 °C
pKa at 25 °C	2.64 - 3.31

2.1 Synthesis of 2,4-D

2,4-D is commonly prepared by the condensation of 2,4-dichlorophenol with monochloroacetic acid in a strongly alkaline medium at moderate temperatures or by the chlorination of phenoxyacetic acid, but this method leads to a product with a high content of 2,4-dichlorophenol and other impurities. Higher reaction temperatures and alkaline conditions during the manufacture of 2,4-D increase the formation of polychlorinated dibenzo-*p*-dioxin (CDD) by-products. One formulation of 2,4-D was found to contain 6.8 μg/kg of 2,3,7,8-

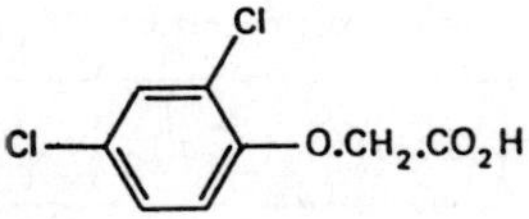

2, 4 – D
2, 4 – dichlorophenoxyacetic acid

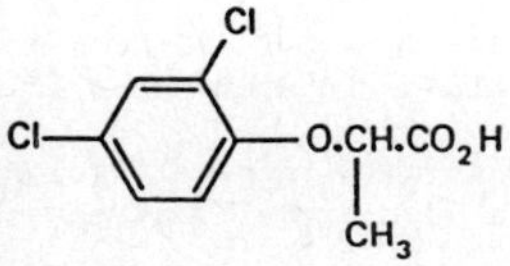

Dichlorprop (2, 4, – DP)
2 – (2, 4 – dichlorophenoxy) propionic acid

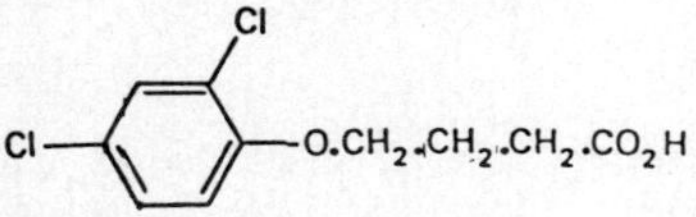

2, 4 – DB
(2, 4 – dichlorophenoxy) butyric acid

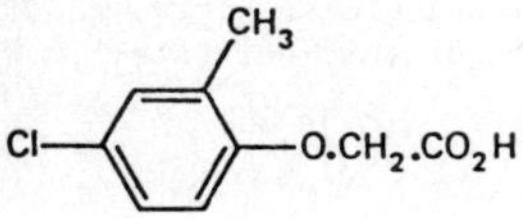

MCPA
4 – chloro – 2 – methylphenoxyacetic acid

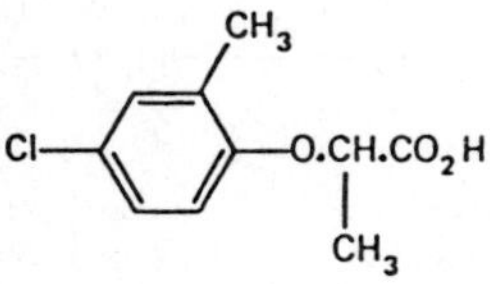

Mecoprop (MCPP)
α– (4 – chloro – 2 – methylphenoxy) propionic acid

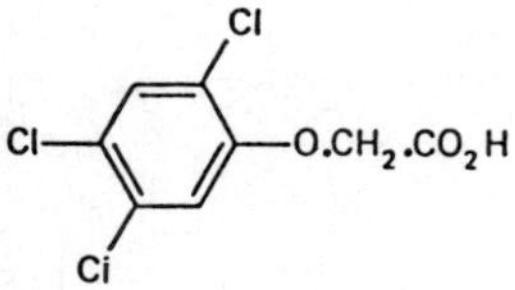

2, 4, 5 – T
2, 4, 5 – trichlorophenoxyacetic acid

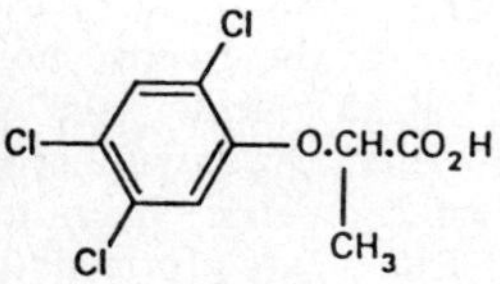

Fenoprop (Silvex, 2, 4, 5 – TP)
2 – (2, 4, 5 – trichlorophenoxy) propionic acid

WHC 83983

Fig. 1. Structures of 2,4-dichlorophenoxyacetic acid (2,4-D) and chemically-related herbicides.

Table 2. Vapour pressure and solubility of 2,4-D salts and esters

Compound	Vapour pressure[a]	Solubility
2,4-D free acid	0.4 mmHg (160 °C)	0.09% in water (25 °C), 85% in acetone (25 °C)
dimethylamine salt		300% in water (20 °C), soluble in acetone
isopropyl ester	1.4×10^{-3} mmHg[b] 4.6×10^{-5} mmHg[b]	insoluble in water, soluble in most organic solvents
butoxyethanol ester (butylethyl ester)	4.5×10^{-6} mmHg[b]	insoluble in water, soluble in most organic solvents
ethylhexyl ester	2.0×10^{-6} mmHg[b]	insoluble in water, soluble in organic solvents
isooctyl ester	2.0×10^{-6} mmHg[b]	insoluble in water, soluble in organic solvents
propyleneglycol butyl ether ester	3.0×10^{-6} mmHg[b]	insoluble in water, soluble in organic solvents
methyl ester	2.3×10^{-3} mmHg[b]	
ethyl ester	1.1×10^{-3} mmHg[b]	
butyl ester	3.97×10^{-4} mmHg[b]	

[a] 1 mmHg = 0.133 kPa.

[b] Vapour pressures of esters were determined at high temperatures by gas-liquid chromatography, and these values are the result of extrapolation to 25 °C. Values vary considerably between authors as a result of this extrapolation; original values at high temperatures agree. Results are presented here as an indication of relative vapour pressure at working temperature. Values from Flint et al. (1968) and Jensen & Schall (1966).

tetrachlorinated dibenzo-*para*-dioxin (Hagenmaier, 1986). In other amine and ester formulations, levels of this dioxin were non-detectable, i.e., < 1 μg/kg (WHO, 1984). The alkali metal salts of 2,4-D are produced by the reaction of 2,4-D with the appropriate metal base. Amine salts are obtained by reacting stoichiometric quantities of amine and 2,4-D in a compatible solvent. Esters are formed by acid-catalysed esterification with azeotropic distillation of water or by direct synthesis in which the appropriate ester of monochloroacetic acid is reacted with dichlorophenol to form the 2,4-D ester.

2.2 Important Chemical Reactions of 2,4-D

Pyrolysis converts various amine salts of 2,4-D to the corresponding amides. Pyrolysis of 2,4-D and its derivatives is likely to produce certain CDD isomers. 2,4-D is readily photodegraded.

2.3 Volatility of 2,4-D Derivatives

2,4-D esters with short-chain alcohols are highly volatile. This influences the effectiveness of their application to target crops, their effects on neighbouring crops, and the degree of contamination of the atmosphere. 2,4-D alkali salts or amine salts are much less volatile than esters, and these products are to be preferred when the use of 2,4-D esters might lead to evaporative 2,4-D losses and to crop damage or damage to the surrounding environment.

Details of technical compositions, impurities, and analytical methods can be found in Environmental Health Criteria 29: 2,4-Dichlorophenoxyacetic acid (WHO, 1984).

3. SOURCES OF ENVIRONMENTAL POLLUTION

The following is a summary of the chapter from Environmental Health Criteria 29: 2,4-Dichlorophenoxyacetic acid (WHO, 1984).

3.1 Production of 2,4-D Herbicides

Comprehensive statistics on 2,4-D herbicide production or use were not available for review. According to the US Department of Agriculture, 3×10^8 kg of total herbicides were used in the USA alone, in 1981. In the past, 10% of the herbicide used was 2,4-D, which would account for a total use in the USA of about 3×10^7 kg. In 1975, an estimated 5×10^6 kg were produced in the United Kingdom. World-wide use of herbicides and annual production, which probably exceeds 5×10^7 kg/year, are increasing.

3.2 Uses

2,4-D alkali or amine salts or esters are used as agricultural herbicides against broad-leaved weeds in cereal crops, as well as on pastures and lawns, in parks, and on golf courses, at rates of about 0.2 to 2.0 kg active ingredient (acid equivalent) per hectare. Esters are also used at rates of up to 6.0 kg (acid equivalent) per hectare to suppress weeds, brush, and deciduous trees along rights-of-way and in conifer plantations and conifer reafforestation areas.

Granular formulations of 2,4-D are used as aquatic herbicides in or along irrigation and other canals, in ponds, and lakes at rates ranging from 1 to 122 kg/ha.

2,4-D products can be used at very low application rates as growth regulators by application of aqueous foliar sprays containing 20 to 40 mg 2,4-D/litre on apple trees to reduce premature fruit-drop, on potato plants to increase the proportion of medium-size tubers or to intensify the tuber skin colour of the red varieties, and in citrus culture to reduce pre-harvest fruit-drop and to increase fruit storage life.

The highly volatile ethyl, isopropyl, and butyl esters are being replaced by low-volatile esters or by amine salts to reduce crop damage resulting from 2,4-D vapour drift, and to decrease atmospheric pollution.

During recent years, the use of 2,4-D and 2,4,5-T in parks, forested recreation, and other areas frequently used by the public, has been reduced in some countries because of increasing concern about possible toxic effects, especially in relation to CDDs.

3.3 Disposal of Wastes

Environmental pollution with 2,4-D may occur as a result of the production and disposal of 2,4-D, or of its by-products, and of industrial effluents. Such pollution will be generally localized to the production site and to areas of waste dumping, and it is likely to be more dispersed if disposal or leaching has occurred into water courses. Disposal of unused 2,4-D in agriculture and washing of equipment may result in localized land pollution and also pollution of water supplies through direct contamination or leaching from soil.

4. UPTAKE, ACCUMULATION, ELIMINATION, AND BIODEGRADATION

Appraisal

2,4-D does not persist in soil because of its rapid degradation.

The physico-chemical properties of 2,4-D acid and its formulations have an important effect on its behaviour in environmental compartments.

The bioavailability to, and uptake by, aquatic and terrestrial organisms is strongly influenced by the organic matter content of soils, microbiological activity, and by environmental conditions such as temperature and pH. Although highly inconsistent, the data on dissipation and bioavailability in various soils demonstrate a marked influence of differences in the texture and mineral composition of the soil (Graham-Bryce, 1972). In aerobic soils, with a high content of organic material, and at high pH values and temperatures, toxic effects are limited because of rapid degradation of 2,4-D.

Uptake is followed by rapid excretion in most organisms. With the exception of some algae, the retention of 2,4-D by organisms in the environment cannot be expected, because of its rapid degradation.

Some microorganisms are capable of utilizing 2,4-D as their sole carbon source. Repeated application to soil stimulates the number of organisms capable of degrading the compound.

4.1 Biodegradation

2,4-D is readily and rapidly degraded in soil. Warm, moist conditions and addition of organic matter stimulate degradation. Autoclaving the soil and inhibiting bacterial metabolism reduce degradation. The kinetics of 2,4-D disappearance suggest that microorganisms are responsible. Particular species of microorganisms, of various types, have been isolated and shown to degrade phenoxyacetic acid herbicides in pure culture. Degradation of the phenoxyacetic acids proceeds by two main pathways. These are via a hydroxyphenoxy acetic acid intermediate or via the corresponding phenol. The literature has been reviewed by the two workers principally responsible for this evidence (Audus, 1960, 1964; Loos, 1969). Some microorganisms are capable of using 2,4-D as their sole carbon source. More often, 2,4-D is co-metabolized with another carbon source. Regular treatment of soil with 2,4-D stimulates the numbers of organisms which are capable of degrading the compound. Treatment with other phenoxy herbicides can also lead to an increase in organisms capable of degrading 2,4-D.

Butler et al. (1975a) exposed 21 species of freshwater algae isolated from natural lake water to 2,4-D butoxyethanol ester, at a concentration of 0.01 mg/litre, and looked for degrading ability. Most of the cultures fully degraded 2,4-D within 2 weeks. A single culture retained 64% of the added 2,4-D, while seven isolates reduced 2,4-D to less than 20% of the amount added. The remaining isolates showed 2,4-D recoveries ranging from 22% to 53%.

Le Van To (1984) isolated six species of microorganisms from soil previously treated with herbicides. These were *Flavobacterium peregrinum*, *Pseudomonas fluorescens*, *Arthrobacter globiformis*, *Brevibacterium sp.*, *Streptomyces viridochromogenes*, and an unidentified *Streptomyces* species. *Flavobacterium* was the most active organism in degrading 2,4-D; degradation of 20 mg/kg of 2,4-D was complete after 20 to 30 days. In a liquid medium, *Flavobacterium* degraded 93.5% of added 2,4-D within 80 h. The time required to degrade half of the 2,4-D added to a sterilized soil along with nutrient was estimated at 3 days. Li-Tse Ou (1984) investigated the breakdown of 2,4-D in two types of soil under dry and moist conditions and at two different temperatures. Numbers of microorganisms degrading 2,4-D were also estimated. Generally, 2,4-D disappeared more rapidly from moist soil; after 14 days of a slow rate of disappearance, however, the removal rate from dry, sandy soil increased. Numbers of organisms degrading 2,4-D were initially much lower in sandy than in clay loams. However, numbers increased rapidly in sandy soils after the addition of the herbicide and, as a result, 2,4-D was eventually degraded more rapidly in sandy than in clay loams. In moist conditions, at 25 °C, the half-life of 2,4-D was 7 days or less, whereas in dry conditions, at 35 °C, it could be as long as 250 days. These latter conditions are unlikely to apply in most natural conditions where 2,4-D is likely to be used.

Rosenberg & Alexander (1980) incubated sewage-sludge bacteria with 2,4-D and found that nearly all of the herbicide had disappeared after 7 days. Subsequent additions of 2,4-D led to destruction of the compound without a lag period; this suggests selection for organisms capable of degrading the compound. Similar results were obtained using bacteria from soil. The time needed for the disappearance of 90% of the added 2,4-D was 14 days with soil inocula. 2,4-D added subsequently was reduced by 70% within 3 to 4 days. Various tropical soils were used in the experiment and all showed a high capacity for degrading 2,4-D. Thompson et al. (1984) determined the persistence of 2,4-D applied at recommended rates in agricultural soils in Canada. In all but one soil, a sandy loam, the concentration had declined by 50% within 7 days. Sattar & Paasivirta (1980) showed slower degradation of 2,4-D in acid soils. It took 6 weeks for 50% of the 2,4-D to disappear from the soil and 7% was still left after 24 weeks. In water-logged soil, there was reduced degradation of the herbicide.

Lewis et al. (1984) studied bacterial breakdown of 2,4-D butoxyethyl ester and the effects of adding various extra components to the medium. The addition of unfiltered, spent fungal medium from which the majority of the fungus had settled out could be either stimulatory or inhibitory to degradation rates of the herbicide; this depended on the particular fungus species cultured in the medium. Further investigation showed that effects were primarily due to differences in pH. Reduction of the pH below 6 inhibited bacterial transformation of the compound. Fungi commonly release large amount of organic acids. The addition of spent fungal medium inhibited the breakdown of 2,4-D ester. Buffering the added fungal medium reduced this inhibitory effect; indeed, some stimulation of breakdown occurred after the addition of buffered, spent medium. The addition of nutrients, or other bacteria which did not transform 2,4-D, stimulated the transformation of the herbicide. The authors consider that the most likely explanation for this phenomenon is induction of other transforming enzymes. With increasing substrate concentration, further enzyme systems are induced in bacteria. The presence of other organisms may stimulate the induction of these other enzymes at lower substrate concentrations than would normally induce them. Increased biomass of transforming bacteria in the presence of competing organisms contributes to increased transformation rates. The nature of the microbial community can, therefore, greatly change the ability of degrading bacteria to transform 2,4-D and other xenobiotics.

O'Connor et al. (1981) found that 2,4-D applied at about 1.5 mg/kg was readily degraded in soil. Adding extra carbon in the form of dried, digested sewage sludge had a short-term effect in enhancing degradation of the compound. Torstensson (1975) measured the half-life of 2,4-D degradation in cultures of soil microorganisms at different pH. In the pH range of 8.5 to 5.0, the half-life changed little, ranging from 5 to 8 days. At pH 4.5, the half-life increased to 21 days and, at pH 4.0, increased further to 41 days.

Lieberman & Alexander (1981) added 2,4-D to inocula of municipal sewage and monitored the biological oxygen depletion (BOD) as a measure of degradation. The herbicide was added to carbon-depleted inocula such that the 2,4-D represented the sole carbon source. Less than 5% of the available oxygen was depleted, indicating poor biodegradation of 2,4-D because of low numbers of organisms capable of degrading the herbicide as their sole carbon source. A separate study showed that 2,4-D was not toxic to microorganisms in sewage.

Fournier (1980) showed that, while 2,4-D treatment increased the numbers of soil microorganisms capable of metabolizing 2,4-D as the sole carbon source and those capable of co-metabolizing the herbicide, this increase was dependent on the concentration of 2,4-D used. At concentrations of 2,4-D between 5 and 50 mg/litre, there was a significant increase in the numbers of organisms metabolizing 2,4-D, and at 5 mg/litre there was a very pronounced increase in organisms co-metabolizing the compound. At much higher (500 mg/litre) or much lower (1.2 μg/litre) 2,4-D concentrations, there was no increase in the numbers of either metabolizing or co-metabolizing organisms.

Sandmann & Loos (1984) estimated the numbers of microorganisms capable of degrading 2,4-D in soils with and without the 'rhizosphere effect' of two plants, African clover *(Trifolium africanum)* and sugar cane *(Saccharum officinarum)*. The 'rhizosphere effect' is a phenomenon which occurs in close association with the roots of plants, where material from the root or the metabolic activity of the root tissue affects the surrounding soil. Particularly high, stimulated populations were associated with sugar cane. A similar effect, but to a lesser degree, was found with clover. In the three sugar cane soils examined, and their corresponding controls, the numbers of organisms were 46 400, 156 000, and 40 700 per g of soil, with rhizospheres, and 178, 1480, and 6170 per g of soil, without rhizospheres, respectively. Seibert et al. (1982) failed to demonstrate a rhizosphere effect on 2,4-D degradation in glasshouse studies using soils with and without maize roots.

Norris & Greiner (1967) investigated the degradation of 2,4-D in forest leaf litter. Litter from either alder, ceanothus, vine maple, bigleaf maple or Douglas fir showed comparable ability to degrade 2,4-D, the recovery of 2,4-D being between 60% and 70% after 15 days of incubation. In a second series of experiments, different formulations of 2,4-D were added to alder litter. About 50% of the free acid of 2,4-D was degraded within 15 days. Triethanolamine salt and two commercial formulations ('solubilized acid' and isooctyl ester) were degraded less than the pure acid. There was between 30% and 40% degradation of these preparations over 15 days.

Nesbitt & Watson (1980) related the degradation rate of 2,4-D in river water to the nutrient levels, sediment load, and dissolved organic carbon content of the water. The addition of sediment or inorganic nutrients increased the rate of 2,4-D degradation, whereas the addition of organisms capable of degrading 2,4-D did not increase the rate of breakdown of the herbicide. This finding indicated that the limiting factor in breakdown of 2,4-D in river water was not numbers of organisms but the nutrient status of the river. The authors noted that in winter, when the river was in peak flow and the water temperature below that for optimum microbial activity, appreciable amounts of the herbicide would be washed into the estuary. An earlier pilot study of seasonal changes in the capacity of river water in Western Australia to degrade 2,4-D (Watson, 1977) indicated clear seasonal differences in both river water concentrations of the herbicide and the degrading capacity of river water. Several rivers were studied and differences were related to the amount of agricultural run-off, the sediment content of the water, river flow, and temperature. Rivers receiving agricultural run-off degraded 2,4-D better than those receiving run-off principally from forests. This was presumed to be the result of the preconditioning of organisms to the herbicide; the investigation corrected for nutrient content of the water which had been previously shown to affect degradation.

Spain & Van Veld (1983) looked at the degrading ability of microbial communities taken from sediment cores from freshwater, estuarine, and marine sites. Some cores were pre-exposed to 2,4-D. Cores from freshwater sites showed increased degradation of 2,4-D after pre-exposure to the compound, whereas estuarine and marine cores did not show this effect. The adaptation of freshwater cores was maximal after 2 weeks and no longer detectable 6 weeks after pre-exposure.

4.2 Uptake and Accumulation by Organisms

Appraisal

Many studies on the accumulation of 2,4-D have used radioactively labelled herbicide and have monitored uptake by simple counting of the label. This fails to take into account that the label could have been removed from the parent molecule by metabolic breakdown. Values for uptake should, therefore, be treated as a maximum possible uptake value for 2,4-D. Such data would not normally be considered acceptable. However, the accumulation of 2,4-D is so low that these data serve to illustrate that little of the herbicide is accumulated.

4.2.1 Laboratory studies

Eliasson (1973) sprayed leaves of 3-year-old aspen *(Populus tremens)* with the butoxyethanol ester of 2,4-D at 0.5 kg acid equivalent/litre. The plants were then kept in an open-sided glasshouse and residues of 2,4-D were monitored. Most of the herbicide remained in, or on, the sprayed leaves. The average residue level was 2300 mg/kg fresh weight 1 day after spraying. This level had fallen to 1300 mg/kg after 37 days and, by day 365, the average residue level was 870 mg/kg. This was a very high application rate and indicates that there is no foliar uptake of 2,4-D by plants.

Glynn et al. (1984) exposed coral *Pocillopora damicornis* to three concentrations of 2,4-D sodium or amine salts at 0.1, 1.0, or 10.0 mg/litre. The maximum concentration of 2,4-D found in coral tissue was 0.137 mg/kg after exposure to the amine salt at 10 mg/litre, but residues were not related to the 2,4-D exposure concentration. The highest bioconcentration factor (BCF) of 1.33 was found after exposure to 0.1 mg/litre of the amine salt of 2,4-D, i.e., the coral contained 1.33 times the concentration of 2,4-D in water.

Metcalf & Sanborn (1975) introduced ^{14}C-labelled 2,4-D into model ecosystems consisting of an alga *Oedogonium,* an aquatic plant *Elodea,* a snail *Physa*, and the mosquito fish *Gambusia.* Total ^{14}C in the water was equivalent to 0.205 mg 2,4-D/litre. The highest BCF was in the alga (26.8, based on measurement of radioactivity). Analysis of all components of the ecosystem for 2,4-D, rather than the radiolabel, revealed none of the parent compound. The BCF, therefore, refers to breakdown products rather than 2,4-D itself. Gile (1983) introduced ^{14}C-labelled 2,4-D, as the butyl ester, into a simulated ryegrass ecosystem. The system consisted of a sandy loam soil, annual ryegrass, several invertebrates, and grey-tailed voles. Voles were introduced 10 days after spraying 2,4-D as a foliar spray at the equivalent of 1 kg/ha. The experiment was terminated after 1 month. Plant material contained an average of 8.9 mg/kg; this was identified as being mostly 2,5-dichloro-4-hydroxyphenoxyacetic acid. Residue levels in animals (based on unidentified ^{14}C residues) ranged from 0.31 mg/kg in snails to 5.28 mg/kg in pillbugs (isopods).

Freitag et al.(1982) measured the bioaccumulation of ^{14}C-2,4-D in an alga *Chlorella fusca* and a fish, the golden orfe. They measured a 24-h static BCF of 6 for the alga, and a 3-day static BCF of <10 for the fish. This measurement was based on radioactivity and, therefore, did not distinguish between the parent compound and its breakdown products.

Schultz (1973) examined uptake and loss of ^{14}C-2,4-D dimethylamine salt by organs of three species of fish (channel catfish, bluegill sunfish, and largemouth bass), exposed to 0.5, 1.0, or 2.0 mg/litre of 2,4-D acid equivalent. After exposure to the highest concentration of 2,4-D dimethylamine salt, there was detectable radioactivity in all organs examined. Bile showed the highest residues of ^{14}C in all three species after 1 week. For the remainder of the exposure period of 12 weeks, there was an increase of radioactivity in other organs and a decrease in the bile. At the end of the exposure period, there was no clear pattern to residue levels of ^{14}C in different organs. These levels ranged from 5.04 mg/kg in bile to 35.5 mg/kg in posterior kidney for the channel catfish. For largemouth bass, the range was from 1.32 mg/kg in muscle to 7.29 mg/kg in liver. For the sunfish, the lowest residue was 24.75 mg/kg in bile and the highest 322.7 mg/kg in the pyloric caeca of the gut. After 84 days exposure to the dimethylamine salt at 2 mg/litre, levels of ^{14}C in the muscle of catfish, bass, and sunfish were equivalent to 0.953, 0.035, and 1.065 mg 2,4-D/kg, respectively. No analysis for 2,4-D itself was carried out. A second study exposed the three fish species for 2 weeks to ^{14}C-2,4-D dimethylamine salt at 1 mg/litre and then for a further 4 weeks to clean water. The disappearance of ^{14}C was measured. Loss of ^{14}C was slow at first but by 4 weeks most tissues had shown a decline in residues. Samples were analysed for 2,4-D but none was detectable, suggesting that the ^{14}C measured was in breakdown products. The values for 2,4-D residues in this and other studies using ^{14}C-labelled material should, therefore, be regarded as overestimates of retained 2,4-D. Uptake of ^{14}C-2,4-D was examined at two different temperatures, 17 °C and 25 °C. The highest residues of ^{14}C detected in fish were equivalent to 0.122 mg 2,4-D/kg, but no 2,4-D could be found after analysis, except in bluegill sunfish after 14 days. Loss of 2,4-D did not, therefore, seem to change with differing temperature over this range. A similar study, at two different water pH values, showed significantly more ^{14}C uptake in all three species at the more acidic pH. Analysis of fish tissues for 2,4-D by gas-liquid chromatography showed non-detectable, or trace, levels in most samples. Only in bluegill sunfish after 7 and 14 days were residues measurable. These 2,4-D residues showed the opposite trend to the ^{14}C results; there was more 2,4-D in fish exposed at the more alkaline pH. The authors suggest that metabolism of the herbicide in the fish is suppressed at alkaline pH.

Sigmon (1979) exposed bluegill sunfish to 2,4-D butyl ethyl ester (3 mg/litre) at three different temperatures, 20, 25, and 30 °C, and measured the tissue content of 2,4-D after 8 days. None of the groups differed from the controls, residues being <0.05 mg/kg.

Bluegill sunfish and channel catfish took up <0.5% of the available ^{14}C when exposed to ^{14}C-2,4-D dimethylamine salt at 2 mg/litre (with 1 litre of water per fish) for 7 days (Sikka et al., 1977). A maximum total ^{14}C concentration in the fish was reached after 24 h and did not change significantly over 14 days. Bluegill sunfish attained a total body concentration of 0.9 mg/kg and catfish 0.2 mg/kg at 24 h. These values were 2,4-D equivalents of ^{14}C measured; the compound was not analyzed directly. When bluegill sunfish were injected intraperitoneally with ^{14}C-2,4-D dimethylamine salt, at dose levels of 1 or 2.5 mg/kg body weight, they excreted 90% of the dose within 6 h of treatment. In a similar experiment, Stalling & Huckins (1978) exposed bluegill sunfish to ^{14}C-2,4-D dimethylamine salt at 2 mg/litre and measured both ^{14}C and 2,4-D in fish and water samples over the following 12 weeks. Radioactivity was detected in tissues and increased over the experimental period, but there was no measurable 2,4-D; the detection limit of the method was 0.1 mg/kg. An *in vivo* intraperitoneal injection of 110 μg of ^{14}C-2,4-D was followed by rapid elimination.

Rodgers & Stalling (1972) measured uptake of ^{14}C-2,4-D butoxy-ethanol ester by three species of fish, which were exposed to either 0.3 or 1.0 mg/litre and sampled over the next 168 h. Some fish were fed and some fasted. Radioactivity in a variety of tissues was determined; the maximum levels were found within 3 h of exposure in fed fish. After this, levels declined over the remaining sampling period, and by the end of the experiment, residues were negligible. The one exception was the gall bladder, which consistently contained more 2,4-D than other tissues. Results were different for fasted fish. In almost all organs of fasted fish, uptake of 2,4-D was slower than for fed fish, although the levels reached were eventually two to five times higher than in fed fish. Analysis of the residues showed that only the liver ever contained the herbicide in the ester form. In all other tissues, only the acid was present.

Shcherbakov & Poluboyarinova (1973) monitored the accumulation of 2,4-D in carp and *Daphnia*. The 2,4-D was added as the butyl ester at concentrations ranging from 0.006 to 5 mg/litre; the recommended usage rate for this ester leads to water concentrations of about 0.5 mg/litre. Analyses of fish tissues were made for both the ester and the acid. The highest BCF for the ester, at 395, was found with fish after a 7-day exposure to 0.5 mg/litre. Acid accumulation was lower than that of the ester. The experiment lasted for 70 days. At day 10 and after, only trace amounts of ester were found in fish. Small amounts of 2,4-D acid were found at day 10, but only trace amounts after day 70. Residues of 2,4-D ester in *Daphnia* varied from 23.9 to 518 mg/kg, according to the exposure concentration.

Two experiments have been carried out on the grey slug *Derocerus reticulatum* by Haque & Ebing (1983) using ^{14}C-labelled 2,4-D acid. The first study, a contact experiment, exposed the slugs to 2,4-D in contaminated soil at 1.1 mg/kg. The body content of 2,4-D in slugs reached equilibrium (0.014 mg/kg) after 15 days; this represented a

BCF of 0.013 based on radioactivity. In the second experiment, slugs were exposed via the food using carrot discs containing 1.1 mg/kg slug body weight per day over 5 days. Residues of ^{14}C in the slugs increased during the feeding period, peaking at 5.5 mg/kg. During the following 7 days, residues were monitored to investigate loss of radioactive material. At the end of the experiment, on day 12, residues were comparable to those at the end of the feeding period. During the course of feeding 2,4-D-contaminated carrots, more than 80% of the ingested dose of radioactivity was excreted rapidly; only 20% was retained. There was no attempt to characterize the ^{14}C residues; these may, therefore, represent either 2,4-D or its breakdown products.

Chickens given a single oral dose of 100, 200, or 300 mg/kg body weight reached maximum plasma levels of 2,4-D of 90, 130, and 250 μg/ml, respectively. Plasma levels in all groups had fallen to 15 μg/ml or less after 24 h. Continuous dosing of chickens at 300 mg/kg per day led to a faster rate of elimination of the daily dose of 2,4-D with time (Bjorklund & Erne, 1966).

4.2.2 Field studies

Cope et al. (1970) treated experimental ponds with 2,4-D propylene glycol butyl ether ester to give water concentrations up to and including 10 mg/litre. No detectable 2,4-D was found in fish exposed to 1 mg/litre or less of the herbicide, but residues were found in bluegill sunfish exposed to 5 or 10 mg/litre. The highest residue (2 mg/kg) was found 1 day after application. Residues were still detectable after 3 days but not subsequently. Vegetation *(Potamogeton nodosus)* and bottom sediment contained residues of 50.0 and 3.0 mg/kg, respectively, 2 days after treatment with the 2,4-D ester at 10 mg/litre. The herbicide was still detectable at 0.1 mg/kg in sediment after 44 days but not thereafter. At 44 days after treatment, there were residues in the plant of 1.2 mg/kg; this amount declined to 0.1 mg/kg after 94 days.

Following the field application of 2,4-D butoxyethanol ester at 22.5 kg/ha, Whitney et al. (1973) measured residues of the herbicide in fish, crustacea, and insect larvae over a 3-week period. The herbicide had been applied to control eurasian water milfoil. Some 2,4-D was taken up by these various species; the highest residue concentration was 0.24 mg/kg in largemouth bass after 8 days. All residues in organisms were below 0.1 mg/kg after 3 weeks. No 2,4-D could be detected in water in 33 samples taken after treatment, the detection limit being 0.10 mg/litre. The highest reported concentration of 2,4-D in mud was 0.65 mg/kg, 10 days after treatment, but in most samples the herbicide level in mud was much lower and in several it was undetectable.

Hoeppel & Westerdahl (1983) treated four areas (10 ha each) of dense water milfoil beds in Lake Seminole, Georgia, with either 2,4-D dimethylamine salt or 2,4-D butoxyethanol ester, at each of two application rates (22.5 or 45 kg/ha). Both formulations were converted to 2,4-D free acid within 24 h. Maximum water concentrations achieved in the high rate (45 kg/ha) areas were 3.6 and 0.68 mg/litre for the dimethylamine salt and butoxyethanol ester, respectively. There was no detectable uptake of 2,4-D into fish in those areas treated with the dimethylamine salt. In the ester-treated areas, 4 out of 24 game fish sampled contained low levels of 2,4-D in muscle (the highest residue being 0.29 mg/kg) and 18 out of 20 gizzard shad contained detectable 2,4-D in muscle (the highest residue being 6.9 mg/kg). No fish sampled more than 13 days after treatment contained detectable 2,4-D.

Schultz & Harman (1974) treated nine experimental ponds with 2,4-D dimethylamine salt at three concentrations: 2.24, 4.48, and 8.96 kg/ha. Samples of water, bottom sediment, and fish were taken over 147 days. Maximum water and sediment concentrations of 2,4-D were 0.692 mg/litre and 0.17 mg/kg, respectively. Of 307 fish sampled, 45 contained detectable residues of 2,4-D. The highest residue measured was in a channel catfish at 1.075 mg/kg 1 day after treatment. All residues in fish after 28 days were less than 0.005 mg/kg; most were undetectable.

Smith & Isom (1967) measured uptake and retention of 2,4-D after treatment of two field sites for control of watermilfoil with the butoxyethanol ester. The first site was treated with a granular formulation at a rate of 112 kg/ha. One bluegill sunfish *(Lepomis macrochirus)* contained 0.15 mg 2,4-D/kg on day 50 after treatment. All other fish, sampled between 72 h and 50 days after treatment, contained less than 0.14 mg/kg, which was the limit of detection. Two samples of several species of mussel, held in cages for 96 h following spraying, showed residues of 0.38 and 0.7 mg/kg. Water levels of 2,4-D reached a peak of 37 mg/litre within 1 h of application and had fallen to less than 1 μg/litre within 8 h. Mud samples contained very variable levels of 2,4-D residues, ranging between 0.14 and 58.8 mg/kg. The highest residue was found 10 months after application. The second site was treated at the lower rate of 45 kg/ha. All fish sampled between 15 days and 9 months after 2,4-D application showed residues of less than 0.14 mg/kg. Mussels sampled between 1 and 42 days after application contained residues ranging between <0.14 and 1.12 mg/kg. Water levels peaked at 157 μg/litre, 1 h after spraying, and mud residues ranged from <0.14 to 33.6 mg/kg.

Coakley et al. (1964) measured residues in organisms at the center of a 0.4-ha field plot sprayed with 2,4-D butoxyethanol ester at a rate of 33.7 kg/ha for watermilfoil control. Two days after application, oysters *(Crassostrea virginica)*, clams *(Mya arenaria)*, fish *(Lepomis gibbus)*, and blue crabs *(Callinectes sapidus)* contained 3.5, 3.7, 0.3, and <0.8 mg/kg, respectively.

In 1971, over 2800 ha in Loxahatchee National Wildlife Refuge were sprayed with the dodecyl-tetradecyl amine salts of 2,4-D at a rate of 4.48 kg/ha. The initial application of 2,4-D was followed by spot treatments of the same formulation and/or the dimethylamine salt of 2,4-D. The highest water concentration (0.037 mg/litre of 2,4-D) was measured 1 day after the initial application. Of 60 fish sampled in the area, 19 had measurable residues of 2,4-D but only three of these were greater than 0.1 mg/kg; the highest recorded residue was 0.162 mg/kg. Breast muscle and liver of a bird, the common Florida gallinule *Gallinula chloropus*, had residues of 0.3 and 0.675 mg/kg, respectively, 1 day after spraying. No residues were found in the bird 4 days after spraying (Schultz & Whitney, 1974).

Plumb et al. (1977) treated sprouting chamise *(Adenostoma fasciculatum)* with the polyethylene glycol butyl ether ester of 2,4-D at a rate of 3.4 kg acid equivalent/ha. A maximum concentration of herbicide (95.2 mg/kg) was found in the plant within 15 min of application. A residue of 3.8 mg 2,4-D/kg remained in, or on, the plants (shoots which had been originally sprayed) 1 year after treatment. When Radosevich & Winterlin (1977) applied the butoxypropyl ester of 2,4-D to a chaparral area at a rate of 4.5 kg/ha, the residues measured in chamise were 221 mg/kg and in grass and forbs 269 mg/kg within 2 h of application. After 30 days, these levels had dropped to 60 mg/kg for chamise and 21 mg/kg for grass and forbs, and, after 360 days, 0.1 mg/kg was present in chamise. Siltanen et al. (1981) monitored residues of 2,4-D in the fruit of bilberries 1 year after the application of 0.25, 0.75, or 2.25 kg/ha acid equivalent. No residues were detected, the limit of detection being 0.05 mg/kg.

Raatikainen et al. (1979), in a controlled field experiment, sprayed cowberry and bilberry with an ester formulation of 2,4-D. Three application rates were used, 0.25, 0.75, and 2.25 kg acid equivalent/ha, and residues of 2,4-D were measured approximately 1 month after application. Thirty-four days after the application of 0.25 kg/ha, residues in cowberry were 0.3 mg/kg. Cowberries exposed to 0.75 or 2.25 kg/ha were analysed after 35 days and contained residues of 1.0 and 3.7 mg/kg, respectively. Bilberries treated with 0.25, 0.75, or 2.25 kg/ha were analysed 29 days later; residues were 0.1, 1.3, and 4.8 mg/kg, respectively.

4.3 Elimination

James (1979) studied tissue distribution of ^{14}C-labelled 2,4-D in the spiny lobster *(Panulirus argus)*. Labelled herbicide was injected into the pericardial sinus and animals were sacrificed at regular intervals. 2,4-D was taken up from the haemolymph, by the green gland, and excreted unchanged, with an overall half-time of about 8 h. Tuey & James (1980), in a similar study, found that the clearance of 2,4-D from haemolymph, via the green gland, was three to five times greater than the rate of metabolism in the hepatopancreas.

Pritchard & James (1979) studied the renal handling of intravenously injected 2,4-D by the winter flounder *(Pseudopleuronectes americanus)*. 2,4-D, at a concentration of 1 μmol/litre of plasma, was actively secreted into the glomerular filtrate of the kidney with clearances of nearly 500 times the glomerular filtration rate. At higher plasma concentrations of between 10 and 60 μmol/litre, a transport maximum of 0.85 μmol/g of kidney per h was observed.

Koschier & Pritchard (1980) reported a similar study using an elasmobranch fish *Squalus acanthias*. They administered 2.5 μmol ^{14}C-2,4-D/kg to the fish intramuscularly and monitored blood and urine ^{14}C levels. Clearance of total 2,4-D was more than 25 times greater than the glomerular filtration rate, indicating that 2,4-D was being actively secreted by the kidney. 2,4-D was eliminated in the urine as a taurine conjugate, this representing about 95% of the excretory products. The plasma contained, primarily, unconjugated 2,4-D (>90%). It seemed, therefore, that 2,4-D was conjugated with taurine before being excreted in the urine. Guarino et al. (1977), in a similar study on the dogfish *Squalus*, also found that 2,4-D was extensively conjugated to taurine (>90%) and was eliminated predominantly via the urine; 70% of the administered dose appeared in the urine within 4 to 6 days. The highest tissue concentration of 2,4-D (14.5 mg/kg) was found in the kidney after 4 h. Plasma elimination was rapid, with a half-time of 44 min; similarly rapid clearance was seen from the kidney. Half-time estimates for muscle and liver were 2 to 3 days and 5 days, respectively.

5. TOXICITY TO MICROORGANISMS

Appraisal

In general 2,4-D is relatively non-toxic to water and soil microorganisms at recommended field application rates.

No effect of 2,4-D was recorded on 17 genera of freshwater and two genera of marine algae at concentrations up to 222 mg/litre.

No effect of 2,4-D was observed on respiration of either sandy loam or clay loam soils at concentrations up to 200 mg/kg.

N-fixation by aquatic algae is affected at high concentrations of 2,4-D acid (400 mg/litre). An effect of 2,4-D esters on N-fixation occurs from a concentration of 36 mg/litre upwards. N-fixing algae in topsoils appear to be more vulnerable to 2,4-D acid than other algal species. The Cyanobacteria (blue-green algae) are important as the major N_2 source in tropical ponds and soils.

In the range of 25.2 to 50.4 mg/litre, 2,4-D was inhibitory to all types of soil fungi.

Cell division was reduced in a green alga by 2,4-D at 20 mg/litre and stopped at 50 mg/litre. No effect was observed on a natural phytoplankton community after exposure to 2,4-D at 1 mg/litre. However, exposure to esters of 2,4-D reduced productivity in these organisms.

5.1 Aquatic Microorganisms

Hawxby et al. (1977) exposed cultures of three algae (*Chlorella pyrenoidosa, Chlorococcum sp.,* and *Lyngbya sp.,)* and one cyanobacterium (blue-green alga) *(Anabaena variabilis)* to concentrations of 2,4-D in the medium of up to 10 μmol /litre (= 2.21 mg/litre). There was no effect on growth, respiration, or photosynthetic rate.

Gangawane et al. (1980) studied the effects of 2,4-D on growth and heterocyst formation in the nitrogen-fixing cyanobacterium (blue-green alga) *Nostoc*. The organism was cultured for 30 days in 0, 10, 100, 1000, or 1500 mg 2,4-D/litre. Growth was measured by optical density and cells forming heterocysts were counted. Growth was inhibited at both 10 and 100 mg 2,4-D/litre and was eliminated at higher concentrations. There was also reduced heterocyst formation.

Lembi & Coleridge (1975) demonstrated a marked effect of 2,4-D, at concentrations of 110 or 220 mg/litre, on cultures of the green algae *Scenedesmus, Ankistrodesmus,* and *Pediastrum*. After 14 days of culture, the three species under control conditions produced 456×10^2, 634×10^4, and 227 cells or colonies per ml of medium, respectively. Corresponding figures after exposure to 110 mg/litre were 54×10^2, 41×10^4, and 74 cells or colonies per ml, respectively. For both *Scenedesmus* and *Ankistrodesmus*, these values were less than the pre-treatment cell concentrations.

Butler et al. (1975b) exposed unialgal cultures of green algae isolated from Warrior River water to 2,4-D butoxyethanol ester at 0.001, 0.01, 0.1, 1.0, or 4.0 mg/litre. Thirty separate isolates were used. Concentrations less than or equal to 1 mg/litre of the 2,4-D ester did not change the growth pattern of the isolates. However, with a concentration of 4 mg/litre, there was some inhibition of growth, as indicated by a 10% increase in the number of incubates which showed poor growth, or no growth, when compared to controls. Some isolates were unaffected even at this concentration and it can therefore be assumed that 2,4-D butoxyethanol ester might change the species composition of green algae populations.

Bednarz (1981) used 12 pure cultures of green algae and cyanobacteria (blue-green algae) separately and in combination to investigate the effects of 2,4-D acid. Cultures were exposed to concentrations of 2,4-D ranging from 0.001 to 10 mg/litre. Low concentrations of 2,4-D stimulated the growth of most species of algae, whereas high concentrations inhibited growth. Chlorococcal green algae were more sensitive to 2,4-D than were filamentous green algae or cyanobacteria. In further experiments, the authors cultured combinations of sensitive and tolerant species in the same range of 2,4-D concentrations. Tolerant species used in combinations were *Chlorella pyrenoidosa, Dictyosphaerium pulchellum,* and *Scenedesmus quadricaudata.* The first two of these tolerant species reduced the toxicity of 2,4-D to sensitive species in mixed culture. This protective effect was not seen with *Scenedesmus.*

Singh (1974) cultured a filamentous, nitrogen-fixing, cyanobacterium *Cylindrospermum sp.* in concentrations of 2,4-D acid of 0, 100, 300, 400, 500, 600, 800, 1000, or 1200 mg/litre and examined growth and heterocyst formation after 8 days. Both parameters were affected at concentrations higher than 300 mg/litre and cultures were killed at a concentration of 1000 mg/litre. Kapoor & Sharma (1980) exposed cultures of the nitrogen-fixing, filamentous cyanobacterium *Anabaena doliolum* to 2,4-D ethyl ester (as 'Weedone 48' concentrate) at concentrations of 36, 108, 180, 252, or 324 mg/litre. There was a dose-related decrease in cell nitrogen over the whole range of 2,4-D ester exposures. Cell growth was stimulated by lower concentrations of 2,4-D and only inhibited by the highest dose. Tiwari et al. (1984) exposed cultures of a similar nitrogen-fixing, filamentous cyanobacterium *(Anabaena cylindrica)* to 2,4-D acid at concentrations of 0, 100, 500, 700, 1000, or 1500 mg/litre, and examined growth, heterocyst formation, and nitrogen fixation. For all these parameters, there was a stimulatory effect of 2,4-D at 100 mg/litre and a progressive inhibition with higher concentrations. These and similar algae are considered to be a major source of nitrogen in tropical ponds and soils. Das & Singh (1977) cultured the nitrogen-fixing cyanobacterium *Anaebaenopsis raciborskii* in concentrations of 2,4-D acid (sodium salt) of 10, 100, 400, 600, 800, and 1000 mg/litre and measured nitrogen fixation. Control cultures and those exposed at 10 and 100 mg 2,4-D/litre showed no significant differences.

Nitrogen-fixation was inhibited at 400 mg/litre or more and eliminated at 600 mg/litre.

Butler (1963) reported no effect on a natural phytoplankton community of exposure to a 1 mg/litre concentration of 2,4-D (as the acid or dimethylamine salt), or of the dimethylamine salt on pure cultures of *Dunaliella euchlora* or *Platymonas* over 4 h. In a later study (Butler, 1965), natural phytoplankton communities were exposed to esters of 2,4-D. Butoxyethanol ester, propylene glycol butyl ether ester, and ethylhexyl ester reduced productivity (as measured by carbon fixation) by 16%, 44%, and 49%, respectively, at a concentration of 1 mg/litre.

Sarma & Tripathi (1980) monitored cell division in the filamentous green alga *Oedogonium acmandrium* exposed to 2,4-D acid at 1, 5, 10, 20, or 50 mg/litre of culture medium. At up to 10 mg/litre, 2,4-D was found to stimulate cell division; a 168 h exposure to 5 mg/litre increased the incidence of dividing cells by 15% over controls. However, cell division was reduced at 20 mg/litre and stopped at 50 mg/litre. Abnormalities in chromosomes during cell division increased with increasing 2,4-D exposure.

Chai & Chung (1975) examined the effects on growth, photosynthesis, respiration, and chemical composition of exposing cultures of the green alga *Chlorella ellipsoidea* to 2,4-D acid at 22 or 88 mg/litre. At 22 mg/litre, 2,4-D increased growth, photosynthesis, and the cell content of protein and nucleic acids. Carbohydrate content was unchanged. However, at 88 mg/litre, growth was inhibited, photosynthesis was no different from controls, and the cell content of carbohydrate, protein, and nucleic acids was decreased.

Elder et al. (1970) examined the effect of 2,4-D acid on 17 genera of freshwater and two genera of marine algae exposed at 22, 111, or 222 mg/litre. There was no effect on the growth of any of the cultures, even at the highest dose of 2,4-D.

Cultures of the flagellate *Euglena gracilis* were exposed for 24 h to concentrations of 1, 5, 10, 50, or 100 mg/litre or for 7 days to 10, 50, or 100 mg/litre of 2,4-D acid by Poorman (1973). Cultures in 50 and 100 mg 2,4-D/litre yielded 84% and 74%, respectively, relative to controls, over 24 h. Lower concentrations of 2,4-D had a slightly stimulatory effect. After 7 days, there was significant stimulation of yield with 10 mg/litre; the culture yielded 161% compared to a control. There was slight stimulation of growth by 50 mg/litre and a reduction to 78% of control levels with 100 mg/litre.

George et al. (1982) exposed the rotifer *Brachionus calyciflorus* to 2,4-D at 5 mg/litre. Median lethal time (LT_{50}) was 24 h and LT_{100} was 31 h.

5.2 Soil Microorganisms

Pachpande & David (1980) isolated the soil alga *Chlorococcum infusionum* from paddy fields and cultured the organism in the presence of 2,4-D acid at concentrations of 0, 1, 2, 3, 4, and 5 mg/litre.

Growth was estimated as dry weight of algal cells filtered out of the medium. All concentrations of 2,4-D were inhibitory to growth. At the highest 2,4-D concentration of 5 mg/litre, the culture yield was reduced from a control level of 720 mg dry wt/litre of medium to 520 mg/litre.

Cullimore & McCann (1977) applied 2,4-D acid to isolated cores taken from a prairie, loam soil to give approximate concentrations of 1 or 100 mg/kg in the top 2 cm of soil. Soil algal populations were estimated from subsamples of cores taken before treatment and 1, 5, or 20 days after treatment with herbicide. Thirty-one genera of algae were identified, of which five were very sensitive to 2,4-D and were rarely found after treatment. These were *Chlamydomonas, Chlorococcum, Hormidium, Palmella,* and *Ulothrix*. The most resistant genera were *Chlorella, Lyngbya, Nostoc,* and *Hantzschia*; the 'percent sensitivity' of these genera (% of the total number of treatments in which the genus was absent) was 28%, 6%, 22%, and 44%, respectively. The reduction in cell numbers of algae in the top layer of the soil after herbicide treatment was soon offset by an increase in the population of *Chlorella, Stichococcus, Oscillatoria,* and *Spongiochloris,* all of which recovered very rapidly from the herbicide effects. There was, however, an overall reduction in cell numbers of nitrogen-fixing algae.

Mukhopadhyay (1980) measured the bacterial, fungal, and actinomycete populations of soils supporting rice or maize plants which had been treated with various herbicides for weed control. There was no effect of 2,4-D, applied at the recommended rate, either on soil microorganism numbers or on the evolution of carbon dioxide by soil cultures.

Huber et al. (1980) examined the effect of 2,4-D at 0.3, 0.2, or 0.1 mmol/litre (= 66, 44, and 22 mg/litre, respectively) on seven cultures of soil microorganisms. There was no effect on the growth of five of the cultures; these were *Nocardia sp., Pseudomonas fluorescens* in both aerobic and anaerobic culture, *Bacillus subtilis,* and *Ustilago maydis*. There was a small reduction in growth at the highest 2,4-D dose in cultures of *Rhizopus japonicus* and *Aspergillus niger*. 2,4-D had no effect on mycelium growth of three out of four plant pathogenic fungi in culture; *Phytophthora cryptogea* showed reduced mycelial growth at 0.1, 0.2, and 0.3 mmol 2,4-D/litre, but *Fusarium oxysporum, Alternia radicina,* and *Rhizoctonia solani* were unaffected.

Moubasher et al. (1981) added 2,4-D at three doses (1.9, 7.6, and 15.2 mg/kg) either to soil or to agar medium inoculated with soil fungi, and the effects on fungal populations were monitored. In soil, at all three doses, 2,4-D stimulated the fungi. When incorporated in the agar medium, 2,4-D was stimulatory to overall fungal growth and to four individual species of fungus at the lowest dose of 6.3 mg/litre, but inhibitory to two other species. At higher doses of 25.2 or 50.4 mg/litre, the herbicide was inhibitory to all fungi.

2,4-D had a significant inhibitory effect on culture yields of the bacterium *Escherichia coli* only at 10^{-3}mol/litre (= 220 mg/litre).

There was no effect at 10^{-4} mol/litre (= 22 mg/litre) (Toure & Stenz, 1977).

Prescot & Olson (1972) added 2,4-D, at doses of 0, 0.1, 1.0, 10, or 100 mg/litre, to cultures of the soil amoeba *Acanthamoeba castellanii* and monitored growth and reproduction. There was a stimulatory effect of 2,4-D at all dose levels; this effect was most marked at the lowest dose and declined with increasing exposure to 2,4-D. The authors suggest that the amoeba may degrade the 2,4-D and utilize it as a carbon source. However, Pons & Pussard (1980) found no effect of 2,4-D (at 28, 54, or 84 mg/litre) on the reproduction of 23 different strains of free-living soil amoebae.

2,4-D, at 10^{-3} mol/litre in cultures of the ascomycete *Neurospora crassa*, stimulated DNA synthesis but had no effect at lower concentrations of 10^{-4} to 10^{6} mol/litre. These concentrations had no significant effect on either RNA or protein (Schroder et al., 1970).

Naguib et al. (1980) measured growth, respiration, and absorption and utilization of sugar and nitrogen in pre-formed fungal mats of *Aspergillus terreus* over 72 h in the presence of 200 mg/litre of 2,4-D. The herbicide inhibited sugar inversion and consequently sugar absorption. It also reduced the incorporation of nitrogen into protein. Respiration was depressed. Growth of the fungus was suppressed and, on a dry weight basis, culture mass was reduced to below the initial level.

Trevors & Starodub (1983) added 2,4-D to sandy loam and clay loam soils and measured both respiration and electron transport system (ETS) activity. ETS was assessed by measuring the capacity of the soil to reduce 2-(*p*-iodophenyl)-3-(*p*-nitrophenyl)-5-phenyl tetrazolium chloride (INT) to iodonitrotetrazolium formazan (INT formazan). The effects of 2,4-D were tested at concentrations of the herbicide in soil of 0, 10, 25, 50, 75, 100, or 200 mg/kg. There was no effect on soil respiration, monitored either as oxygen consumption or carbon dioxide evolution, at any of the concentrations of 2,4-D in either soil. There was similarly no effect on ETS in the sandy loam. However, in the clay loam, there was a progressive inhibition of ETS over the whole range of concentrations of the herbicide. The control soil had an ETS activity of 37.3 μg INT formazan production/g soil, whereas the ETS activity of soil treated with 10 mg 2,4-D/kg was 25.1 μg INT formazan/g, significantly lower than that of the control. The activity was reduced further with increasing concentrations of 2,4-D, until an activity of 16.3 μg INT formazan/g was found at 200 mg 2,4-D/kg.

Deshmukh & Shrikhande (1975) added 2,4-D, at recommended field rates, and at five times the recommended field rates, to two types of soil from India. Both doses of 2,4-D inhibited numbers of *Azobacter* in both soil types, and the high, but not the low, dose of 2,4-D reduced nitrogen fixation in both soils. The same authors (Deshmukh & Shrikhande, 1974) monitored the populations of various microorganisms under the same dosing conditions. 2,4-D stimulated the numbers of actinomycetes throughout the 6-week incubation period at both dose levels. Fungal populations were reduced in the first week of incubation at both dose levels in sandy loam, but only at the higher dose level in clay loam. This reduction in fungal populations persisted until the second week with the high dose in the sandy soil and throughout the incubation period with the high dose in clay soil. There was a temporary (1 week) reduction in total bacterial numbers with both 2,4-D dose levels in the sandy soil and with the higher level in clay soil. Schroder & Pilz (1983) reported that 2,4-D at approximately 10^{-4}mol/litre (= 22 mg/kg) had no long-term effect on soil nitrification.

Welp & Brummer (1985) measured the influence of 2,4-D on the reducing capacity of soil microorganisms, reduction being monitored as the capacity to reduce Fe(III) oxides to soluble Fe(II) ions. They determined no-observed-effect levels (NOEL) of 115 and 95 mg 2,4-D/kg for two different soil types and corresponding EC_{50} values on reduction capacity of 200 and 530 mg 2,4-D/kg soil.

Ruggiero & Radogna (1985) extracted and partially purified soil diphenolase (laccase) from forest soil. This enzyme, which exists free in the soil, plays an important role in the metabolism of humic materials in soil. Oxygen consumption was monitored during the enzymatic reaction, using either catechol or *p*-phenylenediamine as substrate, and the effect of 2,4-D was investigated. The herbicide inhibited diphenolase activity, and Lineweaver-Burk plots of the data suggested that 2,4-D acts as a non-competitive inhibitor. Apparent K values of 28.7 and 6.0 mol/litre were obtained for catechol and *p*-phenylenediamine, respectively.

6. TOXICITY TO AQUATIC ORGANISMS

6.1 Toxicity to Aquatic Invertebrates

Appraisal

The short-term toxicity data on the effects of 2,4-D free acid, its salts, and esters on aquatic invertebrates is extensive. Ester formulations are more toxic than the free acids or salts. Sensitivity variations exist among species in response to the same formulation. Organisms become more sensitive to 2,4-D when the water temperature increases. Reproductive impairment occurred at concentrations below 0.1 of the short-term toxic levels determined for these formulations.

6.1.1 Short-term toxicity

The short-term toxicity of 2,4-D to aquatic invertebrates is summarized in tables 3 - 5.

Unfortunately, there are few studies where both the free acid (or its salts) and ester preparations have been tested on the same organism under the same conditions. The only organisms for which this applies are the oyster (Butler, 1963; Butler, 1965), the stonefly (Sanders & Cope, 1968), and daphnids and shrimp (Sanders, 1970a). These studies all show that the free acid and its salts are less toxic than ester formulations; for example the free acid is at least 20 times less toxic to the water flea *Daphnia magna* than the least toxic of the esters tested (Sanders, 1970a). Comparing studies carried out by different authors and in different systems also suggests a much greater toxicity of the ester preparations.

Liu & Lee (1975) found that 2,4-D could adversely affect the bay mussel *(Mytilus edulis)* at all stages of its life cycle. The attachment of young mussels to test chamber walls was reduced (data in Table 3). The authors evaluated, in two duplicate experiments, the effects of 2,4-D acid, at concentrations in sea water of 22.8, 45.7, 91.4, and 182.8 mg/litre, on the growth of larval mussels. After 10 days exposure, there was a significant reduction in the growth of larvae exposed to 91.4 mg 2,4-D/litre; larvae were 11.6% smaller than controls. This reduction was found in only one experimental replicate. In both experiments, there was reduced growth after 10 days exposure to 182.8 mg/litre; larvae were 31.9% and 34.9% smaller than controls in the two experiments. Exposure for 20 days at 91.4 mg/litre led to reduced growth in both experiments. All larvae exposed to 182.8 mg/litre died within 12 days, but only in one experimental replicate. Extension of the growth study in the second experiment led to all larvae dying within 22 days of exposure to 182.8 mg/litre and, therefore, failing to undergo metamorphosis. The metamorphosis of larvae exposed from age 30 to 70 days was not affected by 2,4-D at concentrations up to 176 mg/litre.

Table 3. Toxicity of 2,4-D to estuarine or marine invertebrates

Organism	Flow/ stat[a]	Temp (°C)	Salinity (°/oo)	pH	Formulation[c]	Parameter	Water concentration (mg/litre)	Reference
Bay mussel (*Mytilus edulis*)		17.2-18.6	22.9-24.5	6.4-7.8	free acid	96-h LC_{50}	259 (232-289)	Liu & Lee (1975)
		17.2-18.6	22.9-24.5	6.4-7.8	free acid	96-h EC_{50} attachment	262	Liu & Lee (1975)
(trocophore larva)		17.2-18.6	22.9-24.5	6.4-7.8	free acid	48-h EC_{50} normal development	211.7	Liu & Lee (1975)
Eastern oyster (*Crassostrea virginica*)	flow	18	29		butoxyethanol	96-h EC_{50} shell growth	3.75	Butler (1963)
	flow	29	25		isooctyl	96-h EC_{50} shell growth	1.0	Mayer (1987)
	flow	28	25		PGBEE	96-h EC_{50} shell growth	0.055	Mayer (1987)
Copepod (*Nitocra spinipes*)		21	7	7.8	butoxyethanol	96-h LC_{50}	3.1 (2.4-4.1)	Linden et al. (1979)
Brown shrimp (adult) (*Penaeus aztecus*)	flow	30			PGBEE	24-h EC_{50} loss of equilibrium	0.55	Butler (1963)
(adult)	flow	30			PGBEE	48-h EC_{50} loss of equilibrium	0.55	Butler (1963)
(juv.)[b]	stat	26	30		butoxyethanol	48-h LC_{50}	5.6	Mayer (1987)
(adult)	flow	29	26		isooctyl	48-h LC_{50}	0.48	Mayer (1987)

Table 3 (Contd).

Dungeness crab (1st zoel) (*Cancer magister*)	stat	13	25		acid (tech)	96-h LC_{50}	> 10	Caldwell (1977)
(1st instar juv.)[b]	stat	13	25		acid (tech)	96-h LC_{50}	> 100	Caldwell (1977)
Blue crab (juv.)[b] (*Callinectes sapidus*)	stat	24	29		PGBEE	48-h LC_{50}	2.8	Mayer (1987)

a Stat - static conditions (water unchanged for duration of test); flow - flow-through conditions (2,4-D concentration in water continuously maintained.

b juv. - juvenile.

c PGBEE - propylene glycol butyl ethyl ester.

Table 4. Toxicity of 2,4-D to freshwater invertebrates

Organism	Flow/ stat[a]	Temp (°C)	Alkalinity[b]	Hardness[b]	pH	Formulation[d]	Parameter	Water concentration (mg/litre)	Reference
Oligochaete worm (*Lumbriculus variegatus*)	flow	20	30	30	7.8	free acid	48-h LC_{50}	122.2	Bailey & Liu (1980)
	flow	20	30	30	7.8	free acid	96-h LC_{50}	122.2	
Water flea (*Daphnia magna*)	stat	21	260	272	7.4	PGBEE	48-h LC_{50}	0.1	Sanders (1970a)
	stat	21	260	272	7.4	dimethylamine	48-h LC_{50}	4.0	Sanders (1970a)
	stat	17		39	7.2	dimethylamine	48-h LC_{50}	> 100.0	Mayer & Ellersieck(1986)
	stat	21	260	272	7.4	butoxyethanol	48-h LC_{50}	5.6	Sanders (1970a)
	stat	21	260	272	7.4	free acid	48-h LC_{50}	> 100.0	Sanders (1970a)
		20			8.4-8.6	free acid	96-h LC_{50}	417.8	Presing (1981)
		20			8.4-8.6	sodium salt	96-h LC_{50}	932.1	Presing (1981)
Water flea (*Simocephalus serrulatus*)		15.6		44	7.4	PGBEE	48-h LC_{50}	4.9 (4.0-6.7)	Sanders & Cope (1966)
Water flea (*Daphnia pulex*)		15.6				PGBEE	48-h LC_{50}	3.2 (2.4-4.3)	Sanders & Cope (1966)
Copepod (nauplius larva) (*Cyclops vernalis*)	stat	20	31.6	70	6.7	free acid	96-h LC_{50}	8.72 (5.34-11.57)	Robertson (1975)
	stat	20	31.6	70	6.7	alkanolamine	96-h LC_{50}	54.8 (46.45-64.6)	Robertson (1975)

Table 4 (Contd).

Scud	stat	21.1	30		7.1	butoxyethanol	24-h LC_{50}	1.4 (1.1-1.8)	Sanders (1969)
(*Gammarus*	stat	21.1	30		7.1	butoxyethanol	48-h LC_{50}	0.76 (0.51-1.1)	Sanders (1969)
lacustris)	stat	21.1	30		7.1	butoxyethanol	96-h LC_{50}	0.44 (0.31-0.62)	Sanders (1969)
	stat	21.1	30		7.1	PGBEE	24-h LC_{50}	2.1 (1.7-2.5)	Sanders (1969)
	stat	21.1	30		7.1	PGBEE	48-h LC_{50}	1.8 (1.4-2.3)	Sanders (1969)
	stat	21.1	30		7.1	PGBEE	96-h LC_{50}	1.6 (1.2-2.1)	Sanders (1969)
	stat	21.1	30		7.1	isooctyl	24-h LC_{50}	6.8 (4.8-9.7)	Sanders (1969)
	stat	21.1	30		7.1	isooctyl	48-h LC_{50}	4.6 (2.9-7.3)	Sanders (1969)
	stat	21.1	30		7.1	isooctyl	96-h LC_{50}	2.4 (1.9-4.8)	Sanders (1969)
	stat	15.5	260	272	7.4	PGBEE	24-h LC_{50}	4.1 (2.8-5.8)	Sanders (1970a)
	stat	15.5	260	272	7.4	PGBEE	48-h LC_{50}	2.6 (1.7-3.9)	Sanders (1970a)
	stat	15.5	260	272	7.4	PGBEE	96-h LC_{50}	2.5 (1.7-3.7)	Sanders (1970a)
	stat	15.5	260	272	7.4	butoxyethanol	24-h LC_{50}	6.5 (1.0-8.6)	Sanders (1970a)
	stat	15.5	260	272	7.4	butoxyethanol	48-h LC_{50}	5.9 (3.1-11)	Sanders (1970a)
	stat	15.5	260	272	7.4	butoxyethanol	96-h LC_{50}	5.9 (3.1-11)	Sanders (1970a)
Scud	stat	15		272	7.4	dimethylamine	24-h LC_{50}	> 100	Mayer &
(*Gammarus*	stat	15		272	7.4	dimethylamine	96-h LC_{50}	> 100	Ellersieck
fasciatus)									(1986)
Glass shrimp	stat	21	260	272	7.4	PGBEE	48-h LC_{50}	2.7	Sanders (1970a)
(*Palaemonetes*	stat	21	260	272	7.4	dimethylamine	48-h LC_{50}	> 100	Sanders (1970a)
kadiakensis									
	stat	21	260	272	7.4	butoxyethanol	48-h LC_{50}	1.4	Sanders (1970a)
Seed shrimp	stat	21	260	272	7.4	PGBEE	48-h LC_{50}	0.32	Sanders (1970a)
(*Cypridopsis vidua*)	stat	21	260	272	7.4	dimethylamine	48-h LC_{50}	8.0	Sanders (1970a)
	stat	21	260	272	7.4	butoxyethanol	48-h LC_{50}	1.8	Sanders (1970a)

Table 4 (Contd).

Organism	Flow/ stat[a]	Temp (°C)	Alkalinity[b]	Hardness[b]	pH	Formulation[d]	Parameter	Water concentration (mg/litre)	Reference
Freshwater prawn	stat	27		113.9	7.5	sodium salt	24-h LC_{50}	2342	Shukla &
(*Macrobranchium*	stat	27		113.9	7.5	sodium salt	48-h LC_{50}	2309	Omkar (1983)
lamarrei)	stat	27		113.9	7.5	sodium salt	72-h LC_{50}	2267	Shukla &
	stat	27		113.9	7.5	sodium salt	96-h LC_{50}	2224	Omkar (1983)
Freshwater prawn	stat	28		112.7	7.5	sodium salt	24-h LC_{50}	2644	Omkar &
(*Macrobranchium*	stat	28		112.7	7.5	sodium salt	48-h LC_{50}	2536	Shukla (1984)
naso)	stat	28		112.7	7.5	sodium salt	72-h LC_{50}	2435	Omkar &
	stat	28		112.7	7.5	sodium salt	96-h LC_{50}	2397	Shukla (1984)
Freshwater prawn	stat	28		112.7	7.5	sodium salt	24-h LC_{50}	2474	Omkar &
(*Macrobranchium*	stat	28		112.7	7.5	sodium salt	48-h LC_{50}	2381	Shukla (1984)
dayanum)	stat	28		112.7	7.5	sodium salt	72-h LC_{50}	2333	Omkar &
	stat	28		112.7	7.5	sodium salt	96-h LC_{50}	2275	Shukla (1984)
Crayfish	stat	15.5	260	272	7.4	PGBEE	48-h LC_{50}	> 100	Sanders (1970a)
(*Orconectes nais*)	stat	15.5	260	272	7.4	dimethylamine	48-h LC_{50}	> 100	Sanders (1970a)
	stat	15.5	260	272	7.4	butoxyethanol	48-h LC_{50}	> 100	Sanders (1970a)
Red swamp crayfish (imm.)[c] (*Procambarus clarki*)	stat	20		100	8.4	alkanolamine	96-h LC_{50}	1389 (1174-1681)	Cheah et al. (1980)
Sowbug	stat	15.5	260	272	7.4	PGBEE	48-h LC_{50}	2.2	Sanders (1970a)
(*Asellus*	stat	15.5	260	272	7.4	dimethylamine	48-h LC_{50}	> 100	Sanders (1970a)
brevicaudus)	stat	15.5	260	272	7.4	butoxyethanol	48-h LC_{50}	3.2	Sanders (1970a)

Table 4 (Contd).

Stone fly (naiad)	15.5	35		7.1	butoxyethanol	24-h LC_{50}	8.5 (5.7-13)	Sanders &
(*Pteronarcys*	15.5	35		7.1	butoxyethanol	48-h LC_{50}	1.8 (1.5-2.7)	Cope (1968)
californica)	15.5	35		7.1	butoxyethanol	96-h LC_{50}	1.6 (1.3-1.9)	Sanders &
	15.5	35		7.1	acid (tech)	24-h LC_{50}	56 (50-63)	Cope (1968)
	15.5	35		7.1	acid (tech)	48-h LC_{50}	44 (32-59)	Sanders &
	15.5	35		7.1	acid (tech)	96-h LC_{50}	15 (10-22)	Cope (1968)
Midge (larva)	15	78-95	55	7.3-7.8	dimethylamine	24-h LC_{50}	1490	Bunting &
(*Chaoborus*	15	78-95	55	7.3-7.8	dimethylamine	96-h LC_{50}	890 (421-1211)	Robertson
punctipennis)	20	78-95	55	7.3-7.8	dimethylamine	24-h LC_{50}	1124	(1975)

a Stat = static conditions (water unchanged for duration of test); flow = flow-through conditions (2,4-D concentration in water continuously maintained.

b Alkalinity and hardness expressed as mg $CaCO_3$/litre.

c imm. = immature.

d PGBEE = propylene glycol butyl ether ester.

Table 5. Toxicity of 2,4-D to aquatic invertebrates: no observed effect levels

Organism	Flow/ stat[a]	Temp (°C)	Salinity (°/oo)	Alkalinity[b]	Hardness[b]	pH	Formulation[c]	Parameter[d]	Water concentration (mg/litre)	Reference
Eastern oyster (*Crassostrea virginica*)	flow	9	19				free acid	96-h EC_0 shell growth	2.0	Butler (1963)
	flow	30	23				free acid	96-h EC_0 shell growth	2.0	Butler (1963)
	flow	25	28				dimethylamine	96-h EC_0 shell growth	2.0	Butler (1963)
Freshwater oligochaete (*Lumbriculus variegatus*)	flow	20		30	30	7.8	free acid	96-h LC_0	86.7	Bailey & Liu (1980)
Scud (*Gammarus lacustris*)	stat	21.1		30		7.1	dimethylamine	96-h LC_0	100	Sanders (1969)
Grass shrimp (*Palaemonetes pugio*)	stat	20	20				butoxyethanol	24-h LC_0	10	Hansen et al. (1973)
Pink shrimp (*Penaeus duorarum*)							butoxyethanol	48-h LC_0	1.0	Butler (1965)
							PGBEE	48-h LC_0	1.0	Butler (1965)

[a] Stat = static conditions (water unchanged for duration of test); flow = flow-through conditions (2,4-D concentration in water continuously maintained).

[b] Alkalinity and hardness expressed as mg $CaCO_3$/litre.

[c] PGBEE = propylene glycol butyl ether ester.

[d] LC_0 and EC_0 represent the highest dose used which cause no death or no effect, respectively; they are not mathematically determined no-effect levels.

Presing (1981) monitored reproduction over four broods in the water flea *Daphnia magna* exposed to 0, 5, 10, 25, or 50 mg/litre of 'Dikonirt' (sodium salt of 2,4-D). For the first brood, the only significant effect was at 50 mg/litre, whereas the fourth brood was delayed even at 5 or 10 mg/litre. Significant reductions in the average number of young produced for each female were found with the two highest concentrations. Young kept until maturity from each of the tests were themselves exposed to 2,4-D in a repeat experiment. Again there was a significant effect on young produced at 25 and 50 mg/litre.

George et al. (1982) measured lethal times (LT) after exposure of the water flea *Daphnia lumholtzi* to 10 or 20 mg 2,4-D/litre. They reported, for 10 mg/litre, an LT_{50} of 38 h and an LT_{100} of 71 h. For 20 mg/litre, the LT_{50} was 21 h and the LT_{100} was 31 h. Doses of 2,4-D ranging from 0.1 to 50 mg/litre did not affect the behaviour of, or kill, the copepod *Mesocyclops leuckarti* within a 30-day exposure period and so lethal times could not be calculated.

Caldwell (1977) and Caldwell et al. (1979) found the zoeal larva to be the most sensitive life-cycle stage of the Dungeness crab *(Cancer magister)* to the free acid of 2,4-D. Based on the herbicide's toxicity to this stage, the authors suggest a maximum acceptable toxicant level (MATC) of <1 mg/litre. At this concentration, there was no mortality, but there was an effect on moulting.

6.1.2 Behavioural effects

Folmar (1978) tested mayfly nymphs *(Ephemerella walkeri)* in a 'Y'-shaped avoidance maze. A 2,4-D dimethylamine salt solution was run into one arm of the maze and clean water was run into a second arm, both at 400 ml/min. Numbers of nymphs in each arm of the maze were counted after 1 h. No avoidance of 2,4-D was found at concentrations of 10 mg/litre and there was no mortality. At 100 mg/litre there was 70% mortality in the test nymphs but still no avoidance of the herbicide. In a similar experiment using the grass shrimp *(Palaemonetes pugio)* exposed to the butoxyethanol ester of 2,4-D, there was significant avoidance of the herbicide at 1 mg/litre (Hansen et al., 1973).

6.2 Toxicity to Fish

Appraisal

At recommended application rates, the concentration of 2,4-D in water has been estimated to be a maximum of 50 mg/litre. Most applications would lead to water concentrations much lower than this (between 0.1 and 1.0 mg/litre).

LC_{50} values for fish vary considerably. This variation is due to differences in species sensitivity, chemical structure (esters, salts, or free acid), and formulation of the herbicide.

Although the free acid is the physiologically toxic entity, the ester formulations represent a major hazard to fish when used directly as aquatic herbicides (because they are more readily taken up by fish). Amine salt formulations used to control aquatic weeds do not affect adult fish.

The NOEL varies with the species and the formulation: <1 mg/litre (coho salmon) to 50 mg/litre (rainbow trout).

Fish larvae are the most sensitive life stage but are unlikely to be affected under normal usage of the herbicide.

Long-term adverse effects on fish are observed only at concentrations higher than those produced after 2,4-D has been applied at recommended rates.

Few studies are related to the effects of environmental variables, such as temperature and water hardness, on 2,4-D toxicity to fish. Higher temperature possibly increases the toxicity. This might be considered when assessing the safety of 2,4-D to fish during control of aquatic weeds.

Fish detect and avoid 2,4-D only at higher concentrations than those obtained under normal conditions of use.

6.2.1 Effect of formulation on short-term toxicity to fish

The toxicity of different formulations of 2,4-D to fish is summarized in Table 6.

The most comprehensive study on the effects of different formulations of 2,4-D using the same test fish, fingerling bluegill sunfish *(Lepomis macrochirus),* was performed by Hughes & Davis (1963) in static 24-h and 48-h tests. Ester formulations were invariably more toxic than amine salt formulations. Dimethylamine and alkanolamine preparations ranged in toxicity from 166 to 900 mg/litre (LC_{50} in 24-h tests), depending on the commercial preparation used. Although esters were always more toxic than amine salts, there was some variation between different ester formulations (range: 0.9 to 66.3 mg/litre; 24-h LC_{50}). Most of this variation was between different preparations of the least toxic of the esters, the isooctyl ester, which ranged in toxicity from 8.8 to 66.3 mg/litre. All other esters tested produced LC_{50} values of 8 mg/litre or less, the most toxic being the isopropyl with a 24-h LC_{50} of 0.9 mg/litre. The addition of emulsifiers to acid preparations increased 2,4-D toxicity; a formulation with emulsifiers gave an LC_{50} of 8 mg/litre over 24 h, making it comparable to the esters in toxicity. All ester formulations were considered by the authors to present a major hazard to fish when used directly as an aquatic herbicide, whereas the amine salt formulations could be safely used to control aquatic weeds without adversely affecting adult fish (Hughes & Davis, 1963).

A study on a range of ester formulations, using salmonids as test fish, conducted by Finlayson & Verrue (1985), showed that the toxicity for salmonids was similar to that for bluegill sunfish. These authors argue that static tests underestimate the toxicity of 2,4-D esters

Table 6. Toxicity of 2,4-D to fish: effects of different formulations

Organism	Flow/ stat[a]	Temp (°C)	Alkalinity[b]	Hardness[b]	pH	Formulation[c]	Parameter	Water concentration (mg/litre)	Reference
Bluegill sunfish (*Lepomis macrochirus*)	stat	25	40	29	6.9	alkanolamine	24-h LC_{50}	450-900	Hughes & Davis (1963)
	stat	25	40	29	6.9	alkanolamine	48-h LC_{50}	435-840	
	stat	25	40	29	6.9	dimethylamine	24-h LC_{50}	166-542	Hughes & Davis (1963)
	stat	25	40	29	6.9	dimethylamine	48-h LC_{50}	166-458	
	stat	25	40	29	6.9	di-N,N	24-h LC_{50}	1.5	Hughes & Davis (1963)
	stat	25	40	29	6.9	di-N,N	48-h LC_{50}	1.5	
	stat	25	40	29	6.9	2,4-D acid + emulsifiers	24-h LC_{50}	8.0	
	stat	25	40	29	6.9	2,4-D acid + emulsifiers	48-h LC_{50}	8.0	Hughes & Davis (1963)
	stat	25	40	29	6.9	isooctyl ester	24-h LC_{50}	8.8-66.3	Hughes & Davis (1963)
	stat	25	40	29	6.9	isooctyl ester	48-h LC_{50}	8.8-59.7	
	stat	25	40	29	6.9	PGBEE	24-h LC_{50}	2.1	Hughes & Davis (1963)
	stat	25	40	29	6.9	PGBEE	48-h LC_{50}	2.1	
	stat	25	40	29	6.9	butoxyethanol	24-h LC_{50}	2.1	Hughes & Davis (1963)
	stat	25	40	29	6.9	butoxyethanol	48-h LC_{50}	2.1	
	stat	25	40	29	6.9	butyl ester	24-h LC_{50}	1.3	Hughes & Davis (1963)
	stat	25	40	29	6.9	butyl ester	48-h LC_{50}	1.3	
	stat	25	40	29	6.9	mixed butyl + isopropyl esters	24-h LC_{50}	1.7	Hughes & Davis (1963)
	stat	25	40	29	6.9	mixed butyl + isopropyl esters	48-h LC_{50}	1.7	Hughes & Davis (1963)
	stat	25	40	29	6.9	isopropylester	24-h LC_{50}	0.9	Hughes & Davis (1963)
	stat	25	40	29	6.9	isopropylester	48-h LC_{50}	0.8	
	stat	25	40	29	6.9	ethyl ester	24-h LC_{50}	1.4	Hughes & Davis (1963)
	stat	25	40	29	6.9	ethyl ester	48-h LC_{50}	1.4	
Cutthroat trout (juvenile) (*Salmo clarki*)						butyl ester	96-h LC_{50}	0.78 (0.66-0.92)	Woodward (1982)
						PGBEE	96-h LC_{50}	0.77 (0.62-0.96)	Woodward (1982)
						isooctyl ester	96-h LC_{50}	> 50	Woodward (1982)

Table 6 (Contd).

Organism	Flow/ stat[a]	Temp (°C)	Alkalinity[b]	Hardness[b]	pH	Formulation[c]	Parameter	Water concentration (mg/litre)	Reference
Chinook salmon (fry)	flow	9	18	17	7.1	butoxyethanol	96-h LC_{50}	0.315	Finlayson &
(*Oncorhynchus tshawytscha*)	flow	9	18	17	7.1	total 2,4-D	96-h LC_{50}	0.373	Verrue (1985)
(smolts)	flow	15	18	17	7.1	butoxyethanol	96-h LC_{50}	0.375	Finlayson &
	flow	15	18	17	7.1	total 2,4-D	96-h LC_{50}	1.250	Verrue (1985)
	flow	15	18	17	7.1	PGBEE	96-h LC_{50}	0.246	Finlayson &
	flow	15	18	17	7.1	total 2,4-D	96-h LC_{50}	1.117	Verrue (1985)
Rainbow trout (fry)	flow	15	18	17	7.1	butoxyethanol	96-h LC_{50}	0.518	Finlayson &
(*Salmo gairdneri*)	flow	15	18	17	7.1	total 2,4-D	96-h LC_{50}	0.642	Verrue (1985)
	flow	15	18	17	7.1	PGBEE	96-h LC_{50}	0.329	Finlayson &
	flow	15	18	17	7.1	total 2,4-D	96-h LC_{50}	0.514	Verrue (1985)
(smolts)	flow	15	18	17	7.1	butoxyethanol	96-h LC_{50}	0.468	Finlayson &
	flow	15	18	17	7.1	total 2,4-D	96-h LC_{50}	1.338	Verrue (1985)
	flow	15	18	17	7.1	PGBEE	96-h LC_{50}	0.342	Finlayson &
	flow	15	18	17	7.1	total 2,4-D	96-h LC_{50}	1.555	Verrue (1985)
loading factor 4.2 g fish/litre	stat	14	18	17	7.1	butoxyethanol	96-h LC_{50}	1.206	Finlayson & Verrue (1985)
	stat	14	18	17	7.1	total 2,4-D	96-h LC_{50}	1.422	Finlayson &
loading factor 8.8 g fish/litre	stat	15	18	17	7.1	butoxyethanol	96-h LC_{50}	3.689	Verrue (1985)
	stat	15	18	17	7.1	total 2,4-D	96-h LC_{50}	4.487	Finlayson & Verrue (1985)

Table 6 (Contd).

Harlequin fish (*Rasbora heteromorpha*)	flow	20		250	7.2	clay-based pellets	24-h LC_{50}	7000	Alabaster (1969)
	flow	20		250	7.2	resin-based pellets	24-h LC_{50}	3950	Alabaster (1969)
	flow	20		250	7.2	resin-based pellets	48-h LC_{50}	3100	Alabaster (1969)
	flow	20		20	7.2	sodium salt	24-h LC_{50}	1160	Alabaster (1969)
	flow	20		20	7.2	butoxyethyl	24-h LC_{50}	1.0	Alabaster (1969)
	flow	20		20	7.2	butoxyethyl	48-h LC_{50}	1.0	Alabaster (1969)
Rainbow trout (*Salmo gairdneri*)	flow	20		250	7.2	clay-based pellets	24-h LC_{50}	7000	Alabaster (1969)
	flow	20		250	7.2	clay-based pellets	48-h LC_{50}	4800	Alabaster (1969)
	flow	20		250	7.2	resin-based pellets	24-h LC_{50}	3400	Alabaster (1969)
	flow	20		250	7.2	resin-based pellets	48-h LC_{50}	2400	Alabaster (1969)
	flow	20		250	7.2	amine salt	24-h LC_{50}	250	Alabaster (1969)
	flow	20		250	7.2	amine salt	48-h LC_{50}	210	Alabaster (1969)

a Stat = static conditions (water unchanged for duration of test); flow = flow-through conditions (2,4-D concentration in water continuously maintained).

b Alkalinity & hardness expressed as mg $CaCO_3$/litre.

c di-N,N = di-N,N-dimethylcocoamine; PGBEE = propylene glycol butyl ether ester; total 2,4-D = the effect actually observed in the flow-through test; the value which preceeds each "total 2,4-D" value is the calculated effect of the ester alone. The authors determined the degree of hydrolysis of the ester during the course of the test and subtracted the effect due to the free acid produced by this hydrolysis.

because some of the ester is hydrolysed to the less-toxic free acid during the course of even short-term tests. The presence of test fish increases the rate of hydrolysis of 2,4-D esters. In a static test, with two different stocking rates of fish, the apparent toxicity of 2,4-D ester decreased with a greater density of test fish (rainbow trout) because of this enhanced hydrolysis. Results are given in Table 6. In their flow-through tests, results were adjusted to take account of the hydrolysis of ester to 2,4-D acid during the course of the experiment. Two values are given in Table 6 for each test. The first is the calculated effect of the non-hydrolysed ester and the second, entered as 'total 2,4-D', is the observed effect of the mixture of ester and free acid produced by hydrolysis during the course of the experiment. There is as much as a five-fold difference between the two values. Alabaster (1969) examined several formulations of 2,4-D in two species of fish, and found that pelleted herbicide, either as clay-based or resin-based pellets, was the least toxic to fish of any of the formulations tested.

6.2.1.1 Tolerance and potentiation

Chambers et al. (1977) used insecticide-tolerant and insecticide-susceptible populations of mosquito fish and an esterase inhibitor to investigate hydrolytic activation and detoxification of 2,4-D esters. Mosquito fish taken from a wild population which had developed some tolerance to insecticides also showed some slight tolerance to 2,4-D ethyl and butyl esters. This tolerance was most pronounced with the butyl ester, where the 48-h LC_{50} was raised from 0.98 mg/litre in the susceptible fish, to 1.70 mg/litre in the tolerant fish. Further experiments were carried out to find the basis for this tolerance and for the higher toxicity of 2,4-D esters over that of the free acid. The addition of DEF (S,S,S-tributyl phosphorotrithioate), a carboxyl esterase inhibitor, to the toxicity test medium slightly reduced the toxicity of both 2,4-D esters. This finding suggested that the toxic effect of the esters required initial hydrolysis to the acid. The increased toxicity of the esters would result from esters being more readily absorbed into the fish through the gills. The resistance of the insecticide-tolerant population was, at least partially, explained by measuring esterase activity in homogenates of gill and liver from the two fish populations. The tolerant fish hydrolyzed less of both 2,4-D esters than susceptible fish. This effect was most marked with liver homogenates; results from gill homogenates were equivocal. The overall conclusion, put forward by the authors, is that liver hydrolysis 'activates' 2,4-D esters by converting them to the toxic free acid. Hydrolysis in the gill is a 'detoxification' reaction because it reduces the uptake of toxic material. The slight increase in tolerance in the insecticide-resistant mosquito fish is largely the result of decreased activation of the 2,4-D esters by the liver. Antagonism to 2,4-D ester toxicity by DEF is largely the result of inhibition of activation in the liver rather than increased detoxification in either liver or gill (Chambers et al., 1977). Carbaryl, a cholinesterase inhibitor, potentiates the toxicity of 2,4-D butyl ester to brown trout *(Salmo trutta)* (Statham & Lech, 1975). The 4.5-h LC_{50} for 2,4-D butyl ester in static tests was shifted from 30 mg/litre to 11 mg/litre by the addition of carbaryl, at a concentration of 1 mg/litre, to the test water. Carbaryl has no toxicity to the fish at this concentration; the 24-h LC_{50} for carbaryl alone is 6.8 mg/litre. The potentiating effect of carbaryl was itself blocked by atropine, a muscarinic blocker, at a water concentration of 10 mg/litre; the atropine itself was not toxic to the fish at this concentration. In a similar way, carbaryl potentiated the toxicity of several other compounds. The authors suggested a non-specific action, possibly by increasing the uptake of 2,4-D ester from the water. The same potentiation was demonstrated for trout in flow-through tests (Statham & Lech, 1975). Combinations of 2,4-D esters (butyl or propylene glycol butyl) with the herbicide picloram increased the toxicity to fish above the combined toxicity of the individual compounds (Woodward, 1982).

6.2.2 No-observed-effect levels in short-term tests with fish

The NOELs of 2,4-D on fish in short-term tests are summarized in Table 7. Values quoted from Meehan et al. (1974) and from Butler (1965) are based on the lowest dose used in their studies. The values from Birge et al. (1979) are calculated 1% mortality values derived mathematically from a full toxicity curve, and based on young fish exposed to 2,4-D from shortly after fertilization of the eggs until 4 days after hatching. The differing exposure times for the three species tested is due to differences in time to hatching. The variation in these data is, therefore, partly due to species differences in sensitivity to the compound and partly due to exposure times. Newly hatched young fish are more sensitive to 2,4-D than unhatched embryos (see section 6.2.4).

6.2.3 Species differences in short-term toxicity to fish

Variation in the toxicity of 2,4-D to fish with species is summarized in Table 8. Of a range of fish species, examined in the same test conditions, using 2,4-D as the free acid, by Rehwoldt et al. (1977), the most sensitive was the white perch *(Roccus americanus)* with a 24 h LC_{50} of 55 mg/litre, and the least sensitive was the eel *Anguilla rostrata* with an LC_{50} of 427 mg/litre. The grass carp, often used together with herbicides to control aquatic vegetation, was the least sensitive of all species examined, with a 24 h LC_{50} for an amine salt formulation of 3080 mg/litre (Tooby et al., 1980).

6.2.4 Toxicity to early life-stages of fish

Short-term studies on the toxicity of 2,4-D to early life-stages of fish are summarized in Table 9.

Table 7. Toxicity of 2,4-D to fish: no-observed-effect levels

Organism	Flow/stat[a]	Temp (°C)	Alkalinity[b]	Hardness[b]	pH	Formulation[c]	Parameter	Water concentration (mg/litre)	Reference
Pink salmon (fry)	stat	10		10-34		free acid	96-h LC_0	< 1	Meehan
(*Oncorhynchus*	stat	10		10-34		butyl ester	96-h LC_0	< 1	et al.
gorbuscha	stat	10		10-34		isooctyl	96-h LC_0	< 1	(1974)
Chum salmon (fry)	stat	10		10-34		free acid	96-h LC_0	10	Meehan
(*Oncorhynchus keto*)	stat	10		10-34		butyl ester	96-h LC_0	< 1	et al.
	stat	10		10-34		isooctyl	96-h LC_0	1	(1974)
Coho salmon (fry)	stat	10		10-34		free acid	96-h LC_0	10	Meehan
(*Oncorhynchus kisutch*)	stat	10		10-34		butyl ester	96-h LC_0	< 1	et al.
	stat	10		10-34		isooctyl	96-h LC_0	1	(1974)
Sockeye salmon (smolts)	stat	10		10-34		free acid	96-h LC_0	10	Meehan
(*Oncorhynchus nerka*)	stat	10		10-34		butyl ester	96-h LC_0	< 1	et al.
	stat	10		10-34		isooctyl	96-h LC_0	< 1	(1974)
Alaska coho salmon	stat	10		10-34		free acid	96-h LC_0	50	Meehan
(fingerlings)	stat	10		10-34		butyl ester	96-h LC_0	< 1	et al. (1974)
(*Oncorhynchus kisutch*)	stat	10		10-34		isooctyl	96-h LC_0	< 1	Meehan
	stat	10		10-34		PGBEE	96-h LC_0	< 1	et al. (1974)
Oregon coho salmon	stat	10		10-34		free acid	96-h LC_0	10	Meehan
(fingerlings)	stat	10		10-34		butyl ester	96-h LC_0	< 1	et al.
(*Oncorhynchus kisutch*)	stat	10		10-34		isooctyl	96-h LC_0	10	(1974)
Dolly Varden	stat	10		10-34		free acid	96-h LC_0	50	Meehan
(fingerling)	stat	10		10-34		butyl ester	96-h LC_0	< 1	et al.
(*Salvelinus malma*)	stat	10		10-34		isooctyl	96-h LC_0	10	(1974)
Rainbow trout	stat	10		10-34		free acid	96-h LC_0	50	Meehan
(fingerling)	stat	10		10-34		butyl ester	96-h LC_0	< 1	et al. (1974)
(*Salmo gairdneri*)									

Table 7 (Contd).

Spot (*Leistomus xanthurus*)	stat					free acid	48-h LC_0	50	Butler (1965)
Longnose killifish (*Fundulus similis*)	stat					dimethylamine	48-h LC_0	15	Butler (1965)
Mullet (*Mugil cephalus*)	stat					ethylhexyl	48-h LC_0	10	Butler (1965)
White mullet (juvenile) (*Mugil curema*)	flow		sea water			free acid	48-h LC_0	50.0	Butler (1963)
Goldfish (*Carassius auratus*)	flow	18.2-25.8	66.7	53.3	7.84	pot. salt	8-day LC_1	8.2 (2.7-15.0)	Birge et al. (1979)[d]
	flow	18.2-25.8	65.3	197.5	7.78	pot. salt	8-day LC_1	8.9 (3.8-14.6)	Birge et al. (1979)[d]
Largemouth bass (*Micropterus salmoides*)	flow	18.2-25.8	66.7	53.5	7.84	pot. salt	7.5-day LC_1	13.1 (4.4-21-9)	Birge et al. (1979)[d]
	flow	18.2-25.8	65.3	197.5	7.78	pot. salt	7.5-day LC_1	3.2 (1.2-6.0)	Birge et al. (1979)[d]
Rainbow trout (*Salmo gairdneri*)	flow	12.5-14.5	66.7	53.5	7.84	pot. salt	27-day LC_1	0.032 (0.008-0.084)	Birge et al. (1979)[d]
	flow	12.5-14.5	65.3	197.5	7.78	pot. salt	27-day LC_1	0.022 (0.006-0.055)	Birge et al. (1979)[d]

a Stat = static conditions (water unchanged for duration of test); flow = flow-through conditions (2,4-D concentration in water continuously maintained).
b Alkalinity & hardness expressed as mg $CaCO_3$/litre.
c pot. salt = potassium salt; PGBEE = propylene glycol butyl ether ester.
LC_0 obtained by extrapolation and LC_1 mathematically calculated.
d Birge et al. (1979) exposed fish from four days after hatching.

Table 8. Toxicity of 2,4-D to fish: species variation

Organism	Flow/ stat[a]	Temp (°C)	Alkalinity[b]	Hardness[b]	pH	Formulation[c]	Parameter	Water concentration (mg/litre)	Reference
Striped bass	stat	20		50	7.2	free acid	24-h LC_{50}	85.6	Rehwoldt
(*Morone saxatilis*)	stat	20		50	7.2	free acid	96-h LC_{50}	70.1	et al. (1977)
Banded killifish	stat	20		50	7.2	free acid	24-h LC_{50}	306.2	Rehwoldt
(*Fundulus diaphanus*)	stat	20		50	7.2	free acid	96-h LC_{50}	26.7	et al. (1977)
Pumpkinseed sunfish	stat	20		50	7.2	free acid	24-h LC_{50}	120	Rehwoldt
(*Lepomis gibbosus*)	stat	20		50	7.2	free acid	96-h LC_{50}	94.6	et al. (1977)
White perch	stat	20		50	7.2	free acid	24-h LC_{50}	55.5	Rehwoldt
(*Roccus americanus*)	stat	20		50	7.2	free acid	96-h LC_{50}	40	et al. (1977)
American eel	stat	20		50	7.2	free acid	24-h LC_{50}	427.2	Rehwoldt
(*Anguilla rostrata*)	stat	20		50	7.2	free acid	96-h LC_{50}	300.6	et al. (1977)
Carp	stat	20		50	7.2	free acid	24-h LC_{50}	175.2	Rehwoldt
(*Cyprinus carpio*)	stat	20		50	7.2	free acid	96-h LC_{50}	96.5	et al. (1977)
Guppy	stat	20		50	7.2	free acid	24-h LC_{50}	76.7	Rehwoldt
(*Lebistes reticulata*)	stat	20		50	7.2	free acid	96-h LC_{50}	70.7	et al. (1977
Grass carp (*Ctenopharyngodon idella*)	flow	13		270	8.1	amine salt	24-h LC_{50}	3080 (2622-3618)	Tooby et al. (1980)
	flow	13		270	8.1	amine salt	48-h LC_{50}	2540 (2184-2952)	Tooby et al. (1980)
	flow	13		270	8.1	amine salt	96-h LC_{50}	1313 (1116-1544)	Tooby et al. (1980)
Bleak (*Alburnus alburnus*)	stat	10	15		7.8	butoxyethanol	96-h LC_{50}	3.2-3.7	Linden et al. (1979)

Table 8 (Contd).

Mosquito fish (*Gambusia affinis*)	stat	21-22			amine salt	24-h LC_{50}	500	Johnson (1978)
	stat	21-22			amine salt	48-h LC_{50}	445	Johnson (1978)
	stat	21-22			amine salt	96-h LC_{50}	405	Johnson (1978)
Mullet (*Mugil cephalus*)	stat				sodium salt	24-h LC_{50}	68.0	Tag El-Din et al.(1981)
	stat				sodium salt	96-h LC_{50}	32.0	
Longnosed killifish (*Fundulus similis*)	stat				butoxyethanol	48-h LC_{50}	5.0	Butler (1965)
	stat				PGBEE	48-h LC_{50}	4.5	Butler (1965)
Bluegill sunfish (*Lepomis macrochirus*)	stat	25	19	7.0	dimethylamine	24-h LC_{50}	390	Davis & Hardcastle (1959)
	stat	25	19	7.0	dimethylamine	48-h LC_{50}	375	
Largemouth bass (*Micropterus salmoides*)	stat	25	19	7.0	dimethylamine	24-h LC_{50}	375	Davis & Hardcastle (1959)
	stat	25	19	7.0	dimethylamine	48-h LC_{50}	350	
Punti (*Puntius ticto*)	stat	23.5			ethyl ester	24-h LC_{50}	1.6	Verma et al. (1984)
Medaka (*Oryzias latipes*)					sodium salt	48-h LC_{50}	> 40	Hashimoto & Nishiuchi (1978)
Longnose killifish (juv.) (*Fundulus similis*)	flow		sea water		PGBEE	24-h LC_{50}	5.0	Butler (1963)
	flow		sea water		PGBEE	48-h LC_{50}	4.5	Butler (1963)
	flow		sea water		butoxyethanol	24-h LC_{50}	5.0	Butler (1963)
	flow		sea water		butoxyethanol	48-h LC_{50}	5.0	Butler (1963)

a Stat = static conditions (water unchanged for duration of test); flow = flow-through conditions (2,4-D concentration in water continuously maintained).

b Alkalinity & hardness expressed as mg $CaCO_3$/litre.

c PGBEE = propylene glycol butyl ether ester

Table 9. Toxicity of 2,4-D to fish early life-stages

Organism	Flow/ stat[a]	Temp (°C)	Alkalinity[b]	Hardness[b]	pH	Formulation	Parameter	Water concentration (mg/litre)	Reference
Bleak (embryo) (*Alburnus alburnus*)						sodium salt	12-h LC_{50}	159.4	Biro (1979)
						sodium salt	24-h LC_{50}	129.0	Biro (1979)
						sodium salt	36-h LC_{50}	63.9	Biro (1979)
						sodium salt	48-h LC_{50}	12.9	Biro (1979)
(larvae)						sodium salt	12-h LC_{50}	111.2	Biro (1979)
						sodium salt	24-h LC_{50}	70.6	Biro (1979)
						sodium salt	36-h LC_{50}	62.1	Biro (1979)
						sodium salt	48-h LC_{50}	51.6	Biro (1979)
Goldfish (embryo) (*Carassius auratus*)	flow	18.2-25.8	66.7	53.3	7.84	pot. salt	4-day LC_{50}	> 187	Birge et al. (1979)
	flow	18.2-25.8	65.3	197.5	7.78	pot. salt	4-day LC_{50}	> 201	Birge et al. (1979)
(4-day post-hatch)	flow	18.2-25.8	66.7	53.3	7.84	pot. salt	8-day LC_{50}	133.1 (108.6-174.8)	Birge et al. (1979)
	flow	18.2-25.8	65.3	197.5	7.78	pot. salt	8-day LC_{50}	119.1 (98.5-150.6)	Birge et al. (1979)
Largemouth bass (embryo) (*Micropterus salmoides*)	flow	18.2-25.8	66.7	53.3	7.84	pot. salt	3.5-day LC_{50}	165.4 (130.6-274.1)	Birge et al. (1979)
	flow	18.2-25.8	65.3	197.5	7.78	pot. salt	3.5-day LC_{50}	160.7 (122.9-230.6)	Birge et al. (1979)
(4-day post-hatch)	flow	18.2-25.8	66.7	53.3	7.84	pot. salt	7.5-day LC_{50}	108.6 (92.5-138.4)	Birge et al. (1979)
	flow	18.2-25.8	65.3	197.5	7.78	pot. salt	7.5-day LC_{50}	81.6 (64.8-103.5)	Birge et al. (1979)

Table 9 (Contd).

Rainbow trout (embryo) (*Salmo gairdneri*)	flow	18.2-25.8	66.7	53.3	7.84	pot. salt	23-day LC_{50}	11.0 (7.8-15.1)	Birge et al. (1979)
	flow	18.2-25.8	65.3	197.5	7.78	pot. salt	23-day LC_{50}	4.2 (2.8-5.9)	Birge et al. (1979)
(4-day post-hatch)	flow	18.2-25.8	66.7	53.3	7.84	pot. salt	27-day LC_{50}	11.0 (7.8-15.1)	Birge et al. (1979)
	flow	18.2-25.8	65.3	197.5	7.78	pot. salt	27-day LC_{50}	4.2 (2.8-5.9)	Birge et al. (1979)

a Stat = static conditions (water unchanged for duration of test); flow = flow-through conditions (2,4-D concentration in water continuously maintained).

b Alkalinity & hardness expressed as mg $CaCO_3$/litre.

Pot. salt = potassium salt.

Studies on fish eggs and larvae immediately after hatching have been conducted on few species and mainly with simple salts of 2,4-D. There is little information on the effects of the more toxic esters. 2,4-D is clearly toxic for fish early life-stages, within the likely range of water concentrations which would be found after use of the herbicide to control aquatic weeds.

At the 16 - 32 cell blastomere stage, eggs of the bleak *Alburnus alburnus*, developed more slowly than control eggs when exposed to 2,4-D solutions. After 48-h exposure, mortality reached 68% and 79% in those eggs exposed to 2,4-D at 25 and 50 mg/litre, respectively. Control mortality after 48-h exposure was 37%. After 23 h of development, eggs exposed to 25 mg 2,4-D/litre showed normal development while those exposed to 100 mg/litre showed slower embryogenesis or development halted at the morula-gastrula stage. Free-swimming larvae were more sensitive to 2,4-D than eggs; the rate of survival of embryos in tests lasting for between 12 and 48 h was higher than for larvae. In tests lasting for between 24 and 48 h, at concentrations of 2,4-D above 400 mg/litre, no larvae survived. Embryos showed malformations and reduced mobility at concentrations above 100 mg/litre, and at concentrations of 800 mg/litre or more, embryos were immobile (Biro, 1979).

Birge et al. (1979) examined the effects of 2,4-D, as the potassium salt, on the eggs and larvae of three species of fish. Rainbow trout eggs were the most sensitive, largemouth bass eggs less sensitive, and goldfish eggs extremely tolerant to 2,4-D. In all species tested, the larval stages were more sensitive to the herbicide than were the eggs.

Only one study has examined the effects of 2,4-D esters on newly-hatched fish fry and on fertilized eggs. Unfortunately, this study, by Hiltibran (1967), does not record the full experimental details. Four species of fish were used in the study but complete results were given only for the bluegill sunfish. The most toxic preparations were the propylene glycol butyl ether (PGBE) ester and mixed isopropyl and butyl esters with no-observed-effect levels of 2 and 3 mg/litre, respectively. The dimethylamine salt, ethylhexyl ester, and sodium salt formulations were less toxic with no-observed-effect levels at 40, 50, and 100 mg/litre, respectively.

6.2.5 Long-term toxicity to fish

Chronic exposure of sub-adult fish of a variety of species to 2,4-D for 10 months, at a concentration of 0.1 mg/litre, led to no mortality and to no change in the acute response to 2,4-D. The 24-h LC_{50} for the compound was unchanged after 10 months exposure to sub-lethal doses; i.e., no tolerance developed. One species of fish, the guppy *Lebistes reticulatus*, reproduced in captivity. Reproductive success was compared between control fish and fish breeding in water containing 2,4-D at 0.1 mg/litre over 10 months. The ratio of numbers of offspring of treated and control fish was 1.2 (Rehwoldt et al.,

1977). Mount & Stephan (1967) conducted a 10-month study on fathead minnows *(Pimephales promelas)* exposed to the butoxyethanol ester of 2,4-D at 0, 0.01, 0.04, 0.2, or 0.8 mg/litre water in a flow-through test system. Concentrations of 2,4-D of 0.2 mg/litre or less had no observable effect on growth, survival, or reproductive success of the fish, but the highest concentration tested was toxic to eggs. The highest tested no-observed-effect concentration was approximately 1/45 of the 96-h LC_{50} for this species. Finlayson & Verrue (1985) conducted a chronic egg-to-fry test over 86 days using chinook salmon in 2,4-D butoxyethanol ester solutions ranging up to 118 μg/litre. The mortality of salmon during the alevin to fry period was 4.7% and 47.6% for exposures to 60 and 118 μg/litre, respectively. When compared to controls, the length of salmon at 36 days post hatch was significantly reduced after exposure to 60 and 118 μg/litre. Neither survival nor growth of fry were affected at 2,4-D concentrations of 40 μg/litre or less. This maximum acceptable toxicant concentration (MATC) represents 0.13 and 0.11 of the 96-h LC_{50} values for fry and smolts of this species, respectively.

6.2.6 *Behavioural effects on fish*

Folmar (1976) used rainbow trout fry in an investigation to determine whether fish avoided water contaminated by herbicides Trout fry were initially placed in a 'Y'-shaped maze for 15 min. The dimethylamine salt of 2,4-D, then was added to one arm of the maze and clean water in a second arm, both at flow rates of 400 ml/min. The number of fish in each arm was counted after 1 hour and results tested statistically using a Chi-squared test. At concentrations of 2,4-D of 0.1 mg/litre water (approximately equal to water levels after the use of this preparation as an aquatic herbicide), there was no avoidance of the compound; the numbers of fish in each arm of the maze were equal. However, at concentrations of 1.0 or 10.0 mg/litre of 2,4-D, there was significant avoidance of the chemical.

Hidaka et al. (1984) conducted a similar study using medakas *Oryzias latipes*, and tested a wide range of doses for a variety of pesticides and herbicides. In all cases, the fish avoided the chemical in a dose-related manner, but only over a limited range of concentrations. Above this range, presumably because of the onset of toxic effects, there was a dose-related decrease in avoidance response. The authors calculated two values from these chevron-shaped graphs, the avoidance response EC_{65} taken from the increasing curve (AR_{65}) and the DAR_{60} taken from the decreasing curve. For 2,4-D, the value for AR_{65} was 177 (171-182; 95% confidence limits) μg/litre, and for DAR_{60} was 288 (245-338) μg/litre. Compared to other chemicals commonly used or found in water, 2,4-D has a high threshold of detection by fish as indicated by the high avoidance threshold.

Rand & Barthalmus (1980) exposed goldfish to 2,4-D at 20 mg/litre (10% of the 96-h LC_{50} for this species) at different stages during the training period for conditioning the fish to avoid electric shocks.

They found that the herbicide had no effect when given for 24 h on the 9th day of conditioning but was effective in changing the magnitude of the avoidance response when given for the first 24 h of conditioning. Fish exposed for 2 weeks showed significant differences in the pattern, rate of acquisition, and maintenance of the avoidance baseline. Behavioural differences persisted into the post-exposure period in fish exposed to 2,4-D for 2 weeks. The authors point out that short-term toxicity tests do not examine subtle behavioural effects which could be of considerable importance in the wild.

Dodson & Mayfield (1979) assessed the effect of 2,4-D on the "reotaxic response" of rainbow trout, that is their tendency to swim upstream to compensate for flowing water. There was a dose-related effect of 2,4-D at concentrations between 0 and 7 mg/litre. Above this concentration range, there was a fall of more than 50% in the frequency of positive reotaxic response to a revolving drum marked in alternate light and dark stripes and an increase in the frequency of 'no-response'. The authors state that, at "realistic concentrations" of 2,4-D in water, there would be a tendency for fish to be moved downstream because of a reduced reotaxic response.

6.2.7 Effects of environmental variables on toxicity to fish

The toxicity of 2,4-D to fish is related to season. Vardia & Durve (1981) obtained different values for 96-h LC_{50} for the carp *Cyprinus carpio* at different times of the year. Water characteristics did not differ, except for temperature which varied from 39 °C in May, its highest value, to 17 °C in February, its lowest. There was a positive correlation between temperature and toxicity. At 39 °C, the 96-h LC_{50} was 5.6 mg/litre, and, at 17 °C, the LC_{50} was 40.83 mg/litre. As the authors point out, the temperature of the water must be borne in mind when assessing the safety of this compound to fish during control of aquatic weeds. The effect may be one of season rather than temperature; the physiology of the fish also changes throughout the year.

There is some effect of water hardness and pH on 2,4-D toxicity to fish but this is very dependent on the species of test fish (Birge et al., 1979). This effect has not been studied systematically.

6.2.8 Special studies on fish

Chronic exposure of carp to sub-lethal concentrations of 2,4-D (5 mg/litre) led to ultrastructural changes in the liver of the fish (Benedeczky et al., 1984). After 2 months, there was detectable swelling of mitochondria and loss of cristae. There were also large numbers of inclusions in the cytoplasm, interpreted by the authors as bile pigments. Their presence in the bile canaliculi indicated the onset of cholestasis (reduction in bile flow). After 3, 4, or 5 months of exposure, the cholestasis was pronounced with cholesterin crystals appearing as cytoplasmic inclusions. Later, in the 6th month of exposure, there were endoplasmic reticulum changes indicative of changed protein synthesis.

Oxygen consumption by bluegill sunfish was not affected by 2,4-D at a concentration of 3 mg/litre water (Sigmon, 1979).

2,4-D at 10^{-4} mol/litre of medium did not affect the activity of Na/K-ATPase in microsomes from trout gill (Davis et al., 1972).

Verma et al. (1984) detected effects on pituitary and pineal histology in punti *Puntius ticto* exposed for 96 h to 1 mg/litre of Weedone (ethyl ester of 2,4-D). There was a significant effect on cell size of acidophilic pituitary cells but a much more marked effect on cyanophils. Pineal epithelium cell height was significantly greater in exposed fish.

6.3 Toxicity to Amphibians

Appraisal

Amphibian larvae are generally tolerant to amine salts of 2,4-D; the 96-h LC_{50} values exceed 100 mg/litre. Of the species tested, only one was sensitive.

No information is available on reproductive development and differentiation or on tissue levels.

The toxicity of 2,4-D to amphibians is summarized in Table 10. Tadpoles of the Indian toad are particularly susceptible to the compound (Vardia et al., 1984). Lhoste & Roth (1946) showed that 2,4-D, at 5 g/litre, prevented development of the eggs of the common frog (Rana temporaria). At doses between 500 mg/litre and 4 g/litre, there was some development which decreased with increasing dose. These levels are far higher than those likely to be encountered in the environment.

Table 10. Toxicity of 2,4-D to amphibians

Organism	Flow/ stat[a]	Temp (°C)	Alkali- nity[b]	Hard- ness[b]	pH	Formulation	Parameter	Water concentration (mg/litre)	Reference
Chorus frog (tadpole)	stat	15.5	30		7.1	dimethylamine	24-h LC_{50}	> 100	Sanders
(*Pseudacris triseriata*)	stat	15.5	30		7.1	dimethylamine	96-h LC_{50}	> 100	(1970b)
Indian toad (tadpole)		25	210	220	8.3	free acid	24-h LC_{50}	13.77 (11.81-16.05)	Vardia et al. (1984)
(*Bufo melanostictus*)		25	210	220	8.3	free acid	48-h LC_{50}	9.03 (8.23-9.91)	Vardia et al. (1984)
		25	210	220	8.3	free acid	96-h LC_{50}	8.05 (7.29-8.81)	Vardia et al. (1984)
Frog (tadpole)	stat	21-22				amine salt	24-h LC_{50}	255	Johnson (1976)
(*Adelotus brevis*)	stat	21-22				amine salt	48-h LC_{50}	228	Johnson (1976)
	stat	21-22				amine salt	96-h LC_{50}	200	Johnson (1976)
Frog (tadpole)	stat	21-22				amine salt	24-h LC_{50}	321	Johnson (1976)
(*Limnodynastes peroni*)	stat	21-22				amine salt	48-h LC_{50}	300	Johnson (1976)
	stat	21-22				amine salt	96-h LC_{50}	287	Johnson (1976)
Toad (tadpole)	stat	21-22				amine salt	24-h LC_{50}	346	Johnson (1976)
(*Bufo marinus*)	stat	21-22				amine salt	48-h LC_{50}	333	Johnson (1976)
	stat	21-22				amine salt	96-h LC_{50}	288	Johnson (1976)
Common frog (tadpole) (*Rana temporaria*)		17-29				free acid	48-h LC_0	50	Cooke (1972)

[a] Stat = static conditions (water unchanged for duration of test); flow = flow-through conditions (2,4-D concentration in water continuously maintained).

[b] Alkalinity and hardness expressed as mg $CaCO_3$/litre.

7. TOXICITY TO TERRESTRIAL ORGANISMS

Appraisal

For terrestrial application, 2,4-D is usually used in the form of the less volatile, longer-chain esters to reduce drift damage of sprays to broad-leaved crop plants. The herbicide is used on cereal crops and on rangeland against broad-leaved weeds and also in forestry. Thus, insects of all kinds will be exposed to 2,4-D. Birds' eggs in the nest are more likely to be exposed than the adult birds, though adult exposure remains possible, particularly for sitting birds. Ground-nesting species will be exposed from the use of 2,4-D on rangeland and cereals; tree-nesting species will be exposed in forests.

The insecticidal action of 2,4-D is low enough that the compound does not represent a hazard to beneficial insects; there is an adequate safety margin with usage at recommended levels.

Although there is some disagreement in the literature about the toxicity of 2,4-D to birds' eggs, the low uptake of the material through the egg shell suggests that exposure would not affect hatching in normal use of the compound. Adult birds are not affected by short-term exposure to 2,4-D. The likelihood of prolonged exposure of either adult birds or eggs to high levels of 2,4-D is small.

7.1 Toxicity to Terrestrial Invertebrates

Appraisal

Based on the widespread use of 2,4-D and its formulations, insects of many kinds could be exposed to the material. Although the compounds are generally classified as non-toxic for beneficial insects, such as honey bees and natural enemies of pests, some adverse effects have been reported on the early life-stages and adults of some insects.

Esters are less toxic to insects than are salts or the free acid.

Feeding studies dosing worker honey bees (Apis mellifera) with 2,4-D salts in sucrose syrup have generated two estimates of 24-h LC_{50}: 104 and 115 μg/bee (Jones & Connell, 1954; Beran & Neururer, 1955). Morton et al. (1972) fed 2,4-D acid to honey bees in 60% sucrose syrup at 10, 100, or 1000 mg/litre and monitored half-time, i.e., the time taken for 50% of the bees in a cage to die. The half-time was significantly longer than that of controls for the two lowest doses (37.2 days at 10 mg/litre and 40.4 days at 100 mg/litre, compared to a control value of 33.4 days), but was significantly reduced at 1000 mg/litre (18.6 days). The butoxyethanol and isooctyl (commercial formulation) esters of 2,4-D had no effect on survival times at 10, 100, or 1000 mg/litre of syrup when fed under the same conditions as the acid. The dimethylamine salt of 2,4-D (commercial formulation) had no effect at 10 or 100 mg/litre, but did shorten the half-time at 1000 mg/litre.

The effects of 2,4-D on beneficial coccinellid larvae were studied by Adams (1960). Larvae were sprayed with a preparation of mixed amine salts of 2,4-D, at a rate equivalent to 0.56 kg acid equivalent/ha, at different stages of their development 1, 3, 6, 9, or 12 days after hatching. There was a lengthening of the development period when the larvae were treated on days 3, 6, 9, or 12 but no effect when they were sprayed on the first day after hatching. Mortality before pupation was more than doubled in all treated groups, but mortality during pupation was not different from that of controls.

Trumble & Kok (1980) dosed adult thistle-rosette weevils *(Ceuthorhynchidius horridus),* which are used for the biological control of musk thistle, with 2,4-D amine salt at five dose levels between 0.17 and 147.8 kg/ha. No significant mortality was observed, up to 175 days after treatment, at doses up to and including 1.68 kg/ha. At 16.8 and 84 kg/ha, there was significantly increased mortality after day 3 post-treatment. At the highest dose level of 147.8 kg/ha, mortality was increased both on day 3 and subsequently. Five-day LC_{50} values for males and females were calculated at 70.2 and 61.4 kg/ha, respectively. This is 41.8 times the recommended application rate of 2,4-D for males and 36.6 for females. Riviere (1976) reared the European cockroach *(Blatella germanica)* on food containing 1000 mg/kg and reported negligible effects on reproduction.

Gall & Dogger (1967) wetted wheat plants with a 0.3% solution of mixed isopropyl and butyl esters of 2,4-D and exposed the plants to females of the wheat stem sawfly *(Cephus cinctus).* Spraying of the wheat plants was performed at different times relative to oviposition; times were 7 days prior to oviposition, at the time of oviposition, or 7, 14, or 21 days after oviposition. Eggs took about 7 days to hatch. The highest larval mortality (96.4%) occurred after spraying at the time of egg laying. The effectiveness of the 2,4-D in killing larvae decreased with later exposure times. For plants sprayed 7, 14, and 21 days after oviposition, larval mortalities were 68.1%, 60.8%, and 37%, respectively, compared to a control mortality of 30%. When plants were sprayed 7 days before egg laying, larval mortality was 46.9%. Adult flies were not affected by 2,4-D spray.

Muller (1971) exposed beetles (Carabidae) to sand dosed with 2,4-D at 0.2, 1.0, or 2.0 g/m^2. Two species, *Bembidion femoratum* and *B. ustulatum,* both showed more than 50% mortality within 4 days of exposure to 1.0 g 2,4-D/m^2. *B. ustulatum* showed 100% mortality within 10 days when exposed to 1.0 g/m^2 and similar mortality within 4 days when exposed to 2.0 g/m^2. About 20% of the individuals of *B. femoratum* survived the 14-day exposure to both 1.0 and 2.0 g/m^2.

Roberts & Dorough (1984) exposed earthworms to 2,4-D acid sprayed on to filter paper. The papers were wetted and the earthworms exposed to the wetted paper in glass vials. The calculated 48-h LC_{50} value was 61.6 (41 - 92.4; 95% confidence limits) $\mu g/cm^2$.

Rapoport & Cangioli (1963) treated turf with a mixture of (4-chloro-2-methylphenoxy)acetic acid (MCPA), at recommended rates, and the butyl ester of 2,4-D, at 10 times the recommended rate. They reported no effect on soil microarthropods.

7.2 Toxicity to Birds

Appraisal

Birds, and particularly the eggs of ground-nesting species, would be exposed to 2,4-D after spraying. Food items could also be expected to be contaminated by the herbicide. However, most studies on birds and their eggs have been conducted at exposures far higher than could be expected in the field.

LD_{50} values from acute oral and from short-term dietary dosing indicate low toxicity of 2,4-D to birds. In longer-term studies, effects have only been reported at extremely high exposures (for example, kidney effects after dosing in drinking water with concentrations in excess of the solubility of the material). There have been no reported effects on reproductive parameters, even at excessive exposure levels.

A single study reported adverse effects on the embryos of birds' eggs sprayed with 2,4-D. Many studies since have shown no effect on hatchability of eggs and no increased incidence of abnormalities in chicks even after very high exposure to 2,4-D. Other work indicates a very poor penetration of the eggshell by the herbicide. It can only be concluded that after normal, or even after excessive, 2,4-D use, there would be no effect on birds' eggs.

7.2.1 Toxicity to birds' eggs

There have been several studies on the toxicity of various 2,4-D formulations to birds' eggs dosed by different routes.

Spraying eggs of pheasant, red-legged partridge, and grey partridge with the equivalent of 0.55 to 1.1 kg 2,4-D/ha, either before incubation or after 3 days of incubation, was found by Lutz-Ostertag & Lutz (1970) and Lutz & Lutz-Ostertag (1972) to cause embryonic abnormalities. They reported 77% mortality in pheasant eggs, 77% in grey partridge eggs, and 43% in red-legged partridge eggs within the first 19 days of incubation. The eggs were broken open, between days 20 and 22 of incubation for histopathological examination of the embryos; hatching takes place at about 24 days in all species used. A majority of surviving embryos were either wholly or partially paralyzed. Histopathological effects were mainly gonadal. In both male and female embryos, there were abnormalities of the gonad, often severe enough to lead to sterility, and, in the male, abnormal regression of the Mullerian ducts. No control embryos were examined. In a second study by Lutz & Lutz-Ostertag (1973), quail, pheasant, and partridge eggs were sprayed with 2,4-D from two different commercial sources either before incubation or on days 3 or 7 after the start of incubation. The authors reported increased mortality in embryos, reduced hatchability, and increased abnormality in chicks. The only control eggs reported were those in a single incubation of quail. Later work has failed to repeat these results.

Kopischke (1972) found no adverse effects on hatchability and no increase in deformities or later mortality of hatched chicks after spraying eggs of pheasants or chickens with an isooctyl ester formulation of 2,4-D on day 13 of incubation at a dose equivalent to 0.28 kg/ha.

Somers et al. (1974) sprayed chicken eggs, prior to incubation, with concentrations of an amine salt of 2,4-D at up to 15 times the recommended field application rate of approximately 3 kg/ha. There was no effect on hatching success or on the survival of chicks in the period 3 to 4 weeks post hatch. Spraying chicken eggs on days 0, 4, or 18 of incubation with 2,4-D (as a PGBE ester formulation) at up to 10 times the field application rate, had no effect on hatchability or on survival and growth of chicks after hatching (Somers et al., 1978a). Birds hatched from eggs, similarly treated by spraying, showed no significant adverse effects on later reproductive performance (egg laying performance of the females; testis weight or sperm count of males) (Somers et al., 1978b). Hilbig et al. (1976a) found no effect on egg hatch rate or on body weight or malformation rate in chicks, after spraying (at 20 kg/ha) the eggs of Japanese quail, pheasants, and chickens prior to incubation, or 3 days after the start of incubation. In a follow-up study on the reproductive performance of birds hatched from these dosed eggs, Hilbig et al. (1976b) reported no effects on laying capacity, fertility, or hatchability of their eggs.

The effects of 2,4-D dimethylamine salt on the eggs of Japanese quail, grey partridge, and red-legged partridge were studied by Grolleau et al. (1974). Eggs were sprayed with 2,4-D at dose levels equivalent to the recommended application rate (1.2 kg/ha) and at two higher dose levels equivalent to 2.4 and 6 kg/ha. There were no effects on hatching rate, embryonic mortality, or chick mortality in the first month after hatching or on embryonic or chick malformations. In addition, the histopathological examination of partridge thyroids revealed no effects. Residues of 2,4-D were measured in those partridge eggs receiving the highest dose. Very little 2,4-D penetrated the egg shell and the highest residue measured was a total egg content of 19.3 μg (in an 11-g egg, 15 days after treatment). The lack of effect of 2,4-D on sprayed eggs was attributed to the poor penetration of the herbicide. Spittler (1976) found no adverse effects of 2,4-D on hatchability and no increase in chick abnormalities in pheasant or quail eggs sprayed 24 h before hatching with a dose 12 times higher than the recommended application rate. Only at a dose 30 times higher than the recommended rate did hatchability fall by 10% to 15%, relative to controls. No increased incidence of abnormalities was reported at this dose rate.

Hoffman & Albers (1984) immersed mallard eggs for 30 seconds in aqueous emulsions of 2,4-D and calculated an LC_{50} equivalent to a field application rate of 216 (155 - 300) kg/ha. This is 32 times the recommended field application rate. Dunachie & Fletcher (1967) injected chicken eggs with 10, 100, or 200 mg 2,4-D/kg, equivalent to 0.5, 5, or 10 mg/egg, and found reduced hatchability relative to control eggs injected with solvent only. Treated eggs showed 80-90%, 70%, and 50% of the control hatch rate for the three dose rates, respectively. In a similar study, Gyrd-Hansen & Dalgaard-Mikkelsen (1974) found that injecting 1 mg/egg or less of the dimethylamine salt of 2,4-D had no effect. Injections of 2 mg/egg reduced both hatchability of the eggs and survival of hatched chicks. An injected dose of 5 mg/egg reduced the hatching rate to 15% of control levels and there were no surviving chicks after 1 week. There was no successful hatching after an injection of 10 mg/egg. The same authors also dosed eggs by immersion in solutions of 2,4-D for 10 seconds. There was no effect after immersion in a solution of 10 g/litre and only a slight effect after immersion in 50 g/litre. The hatching success and survival of the chicks up to 4 weeks post hatch, after immersion in 50 g/litre, was more than 80% of control values.

7.2.2 Toxicity to birds after short-term and long-term dosing

The toxicity of 2,4-D (given either orally by capsule or in the diet) to birds is summarized in Table 11. The studies reported in this Table include single oral dosing, repeated oral dosing, and dietary tests over 5 to 100 days. Studies lasting 10 days or less show that high dosage (in excess of 1000 mg/kg food) is required to kill birds. 2,4-D is, therefore, of low toxicity to birds.

Haegele & Tucker (1974) dosed egg-laying Japanese quail and mallard a single oral dose of 250 or 1500 mg 2,4-D acid/kg body weight and monitored egg shell thickness. There was a short-term effect; thin-shelled eggs were produced during the first 3 days after dosing. This was considered to be an indirect effect, i.e., the result of reduced food consumption. When Bjorklund & Erne (1966) gave single oral doses of 100, 200, or 300 mg 2,4-D amine/kg body weight to chickens, all clinical and gross pathological findings were negative, with the exception of a single bird showing gastritis after the highest dose. 2,4-D amine was given orally at 300 mg/kg body weight to a second group of chickens each day. One bird died after 5 days and was shown on autopsy to have developed renal and visceral gout. The other birds were killed on days 12 or 24 of dosing. Slight kidney enlargement was seen, and there was an enhancement of the rate of 2,4-D elimination with time.

Bjorn & Northen (1948) orally dosed white-rock chicks on alternate days for a period of 4 weeks (12 doses in total) with an alkanolamine salt formulation of 2,4-D. All chicks weighed approximately 50 g at the beginning of dosing and doses were adjusted for weight gain of the chicks through the dosing period. No effect on weight gain was noted at doses of 2,4-D up to 280 mg acid equivalent/kg body weight. In a further study, single oral doses of up to 380 mg/kg body weight were without effect, but a single oral dose of 765 mg/kg body weight killed all the birds.

Table 11. Toxicity of 2,4-D to birds

Species	Sex[a]	Age	Route	Formulation	Parameter	Concentration[b] (mg/litre)	Reference
Mallard duck	M	4 months	oral	acid (technical)	acute LD_{50}	> 2000	Hudson et al. (1984)
(*Anas platyrhynchos*)	M	3-5 months	oral	sodium salt	acute LD_{50}	> 2050	Hudson et al. (1984)
	M	7 months	oral	amine salt	acute LD_{50}	< 2000	Hudson et al. (1984)
	F	3-5 months	oral	acid (technical)	acute LD_{50}	> 1000	Hudson et al. (1984)
		23 days	diet	butoxyethanol	5-day LC_{50}	> 5000 [c]	Hill et al. (1975)
		17 days	diet	dimethylamine	5-day LC_{50}	> 5000 [c]	Hill et al. (1975)
		young	diet	acetamide	10-day LC_{50}	2500	DeWitt et al. (1963)
		young	diet	acetamide	100-day LC_{50}	> 500	DeWitt et al. (1963)
		adult	diet	acetamide	100-day LC_{50}	> 2500	DeWitt et al. (1963)
		young	diet	dimethylamine	100-day LC_{50}	2500	DeWitt et al. (1963)
		young	diet	butoxyethanol	100-day LC_{50}	5000	DeWitt et al. (1963)
		adult	diet	butoxyethanol	100-day LC_{50}	> 5000	DeWitt et al. (1963)
Japanese quail	M	2 months	oral	acid (technical)	acute LD_{50}	668 (530-842)	Hudson et al. (1984)
(*Coturnix coturnix*		14 days	diet	acetamide	5-day LC_{50}	> 5000 [c]	Hill et al. (1975)
(*japonica*)		12 days	diet	butoxyethanol	5-day LC_{50}	> 5000 [c]	Hill et al. (1975)
		20 days	diet	dimethylamine	5-day LC_{50}	> 5000 [c]	Hill et al. (1975)
Bobwhite quail		23 days	diet	butoxyethanol	5-day LC_{50}	> 5000 [c]	Hill et al. (1975)
(*Colinus virginianus*)		23 days	diet	dimethylamine	5-day LC_{50}	> 5000 [c]	Hill et al. (1975)
		young	diet	acetamide	10-day LC_{50}	2500	DeWitt et al. (1963)
		adult	diet	acetamide	100-day LC_{50}	> 2500	DeWitt et al. (1963)
		young	diet	dimethylamine	10-day LC_{50}	5000	DeWitt et al. (1963)
		young	diet	butoxyethanol	100-day LC_{50}	5000	DeWitt et al. (1963)
		adult	diet	butoxyethanol	100-day LC_{50}	5000	DeWitt et al. (1963)

Table 11 (Contd).

Pheasant	F	3-4 months	oral	acid (technical)	acute LD_{50}	472 (340-654)	Hudson et al. (1984)
(*Phasianus colchicus*)		10 days	diet	butoxyethanol	5-day LC_{50}	> 5000	Hill et al. (1975)
		10 days	diet	dimethylamine	5-day LC_{50}	> 5000 [c]	Hill et al. (1975)
		young	diet	acetamide	10-day LC_{50}	1000	DeWitt et al. (1963)
		adult	diet	acetamide	100-day LC_{50}	> 2500	DeWitt et al. (1963)
		young	diet	dimethylamine	100-day LC_{50}	5000	DeWitt et al. (1963)
		adult	diet	dimethylamine	100-day LC_{50}	> 5000	DeWitt et al. (1963)
		young	diet	butoxyethanol	100-day LC_{50}	5000	DeWitt et al. (1963)
		10 days	diet	butoxyethanol	5-day LC_{17}	5000	Hill et al. (1975)
Chukar partridge (*Alectoris chukar*)	M,F	4 months	oral	acid (technical)	acute LD_{50}	200-400	Hudson et al. (1984)
Rock dove (*Columba livia*)	M,F		oral	acid (technical)	acute LD_{50}	668 (530-842)	Hudson et al. (1984)
Chicken	M,F	21 days	oral	free acid	14-day LD_{50}	541 (358-817)	Rowe & Hymas (1954)
	M,F	21 days	oral	isopropyl	14-day LD_{50}	1420 (1127-1789)	Rowe & Hymas (1954)
	M,F	21 days	oral	mixed butyl esters	14-day LD_{50}	2000 (1350-2960)	Rowe & Hymas (1954)

a M = male; F = female.
b Acute oral doses are given as mg/kg body weight; all other doses are as mg/kg diet.
c Dose level of 5000 mg/kg diet produced no mortality.

Whitehead & Pettigrew (1972a) dosed 28-week-old laying hens daily, by gelatin capsule, with the butoxyethyl ester of 2,4-D at either 6.2 or 18.7 mg acid equivalent/bird for 20 weeks. There were no adverse effects on egg production, egg or yolk weight, egg shell thickness, hatchability, or growth rate of the progeny.

Chickens were given oral doses of 2,4-D amine salt at 100, 250, or 500 mg/kg body weight or PGBE ester at 50, 100, or 250 mg/kg body weight for 10 days in a study by Palmer & Radeleff (1969). Birds given 2,4-D amine salt did not differ from controls in weight gain, even at the highest dose. There was similarly no effect from the lowest dose of the ester. However, a growth rate reduction was seen with the medium dose of the ester (only 19% weight gain relative to a control weight gain of 41%), and at the highest dose, there was complete mortality within 4 days, associated with a weight loss of 13%. In a comparable study, Palmer (1972) dosed chickens with 2,4-D dimethylamine salt at 25 to 500 mg/kg body weight for 10 consecutive days. There were effects on weight gain at doses of 100 mg/kg or more. At 100 mg/kg, the weight gain was 38%, compared to a control value of 57%. At 175, 250, and 375 mg/kg, the weight gain was 30%, similar to the control value. At the highest dose, three out of five treated birds died, and the survivors showed a weight gain of 26%. There was no effect of 2,4-D ethylhexyl ester at 100 mg/kg on weight gain, but at 250 and 500 mg/kg, weight gain was 42% and 36%, respectively, compared to a control value of 59%.

Solomon et al. (1973) studied the effects of 2,4-D acid and two unspecified amine salt formulations of 2,4-D. Pheasants were dosed at weekly intervals, for 17 weeks, with gelatin capsules containing one of the preparations at either 75 or 150 mg/bird. No effects were observed on fertility and there was no increase in the number of abnormal embryos.

Whitehead & Pettigrew (1972b) fed day-old chicken chicks with the butoxyethanol ester of 2,4-D at concentrations up to 7500 mg/kg diet for 3 weeks. Dietary levels up to 1000 mg/kg had no adverse effect, but at 2000 mg/kg diet 2,4-D ester reduced the food consumption and growth rate of the chicks. Although there was no mortality at the higher doses, necropsy of birds sacrificed at the end of the experiment showed swollen kidneys in all birds and some mottling of the spleen. Bjorklund & Erne (1966) fed chickens with the amine salt of 2,4-D at 500 mg/kg diet. One bird died of renal gout after 5 months dosing; autopsy showed hypoplasia (possibly congenital) of the right kidney and hyperplasia of the left kidney. Other birds were killed at 1, 2, 9, or 18 months after dosing began, but there was no consistent pattern to autopsy findings.

Erne & Bjorklund (1970) examined the long-term effects of phenoxyherbicides on chickens. Groups of day-old broiler chicks were given 2,4-D in the drinking water at 1000 mg/litre for up to 7 months. During the dosing period, chickens were sacrificed at regular intervals for autopsy and samples were prepared for electron microscopy. There was decreased food and water intake in dosed birds. The most pronounced effect was on the kidney; there was noticeable kidney enlargement after 14 days of dosing and this increased with time. 2,4-D concentrations in body tissues reached a plateau after 7 days, with the highest residue in kidney tissue. Histologically, the kidney enlargement was shown to be due to hypertrophy of proximal tubule epithelium. These hypertrophied cells, under the electron microscope, were shown to display an increased mitochondrial content and pronounced mitochondrial pleomorphy. The number of microbodies was also increased and nuclear bodies were observed. These findings were stated to reflect alterations in intermediate metabolism in the tubular cells. In an earlier study, using chicks dosed similarly with 1000 mg/litre of drinking water (Bjorklund & Erne, 1966), the birds were followed through to sexual maturity. No significant effects were observed on weight gain, age at sexual maturity, or onset of egg production, but the number of eggs laid was reduced during the first 2 months of egg laying. The number of birds dying during the course of the study did not differ from the control value. Surviving birds were killed and autopsied at intervals of between 2 and 18 months after the onset of egg laying. The primary effect was consistent enlargement of the kidneys.

7.2.3 Special studies on birds

Lundholm & Mathson (1983) studied the effect of 2,4-D on the ATP-dependent Ca^{2+} binding of the particulate fraction of egg shell gland mucosa cells from egg-laying hens. This parameter had been found to be a sensitive indicator of potential shell-thinning effects of chemicals. They calculated a 5-min IC_{50}, for Ca binding inhibition, of 30.7×10^{-8} mmol 2,4-D/litre incubation medium. This makes 2,4-D 13.5 times less effective in this respect than 1,1′-(2,2-dichlorethenylidine)-bis[4-chlorobenzene] (*p-p′*-DDE), the major agent causing eggshell-thinning in birds.

Percutaneous absorption of 2,4-D through the feet of red-winged blackbirds was measured by Rogers et al. (1974). A 24-h exposure to ^{14}C-labelled 2,4-D at 0.01 mmol/litre resulted in a blood concentration of 1.24×10^{-3} mmol 2,4-D/litre.

7.3 Toxicity to Non-laboratory Mammals

Appraisal

Based on the available data, no generalization can be made about the hazard of 2,4-D to mammals in the field. Data on voles indicate that the herbicide poses no hazard.

Cholakis et al. (1982) obtained acute oral LD_{50} estimates for two species of voles by determining mortality 14 days after the administration of a single dose of 2,4-D acid. Values were 2110 (1800 - 2570) and 2100 (1900 - 2390) mg/kg body weight for males and females, respectively, of the prairie vole *(Microtus orchrogaster)*. Values for the grey-tailed vole *(Microtus canicaudus)* were 1200 (955 - 1150) for males and 1310 (1010 - 1790) mg/kg body weight for females.

Skokova (1975) orally dosed 24 male bank voles with 400 to 405 mg/kg body weight (10% of the LD_{50}) daily for 10 or 20 days and examined reproductive parameters. Testis weight, an index of spermatogenesis, and divisions in spermatogonia were all significantly reduced relative to control values. Gile (1983) applied a foliar spray of butyl ester of 2,4-D to a simulated ryegrass ecosystem at 1 kg/ha. The system included voles, which showed a weight loss after exposure to 2,4-D when compared to similar animals in an untreated system. This loss was considered to be the result of protein deficiency.

8. ECOLOGICAL EFFECTS FROM FIELD APPLICATION

Appraisal

No direct toxic effects, acute or long-term, of 2,4-D applications under field conditions on any animals species have been observed thus far.

There are, inevitably, indirect effects resulting from the intended selective herbicidal properties of the compound. These effects would result from the use of any herbicide or from other methods of land management. There will, therefore, be effects for mammals, birds, and insects because of food deprivation, modification of habitat, requirements for nesting, shelter, etc.

The application of 2,4-D appears to present no hazard to the beneficial epigeal arthropod community. Physical cultivation present a greater hazard to sensitive soil arthropods than the use of 2,4-D herbicides.

Oka & Pimental (1976) observed increased numbers of insect pests and increased occurrence of blight infection in maize *(Zea mays)* crops treated with 2,4-D as the triethanolamine salt. The crops had been treated with 2,4-D at 0.14, 0.55, or 4.4 kg/ha; 0.55 kg/ha is the normal rate of application for this crop. The number of aphids increased from 1420 on control untreated plants to 2449 on plants treated at 0.14 kg/ha, 3116 on plants treated at 0.55 kg/ha, and 2023 on plants treated at 4.4 kg/ha. The percentage of plants attacked by the European corn borer *(Ostrinia nubialis)* increased from 63% on controls to 83% and 70% for treatments at 0.14 and 0.55 kg/ha, respectively. Controlled studies also showed an increase in infection with fungal blight. Laboratory investigation of these effects confirmed that the treated maize had higher protein levels than the untreated. This was thought to be the reason for the increased success of the pests.

Everts et al. (1986) investigated the effects of various pesticides on soil arthropods. Spiders were found to be a sensitive indicator of effect. No side-effects of the use of 2,4-D amine were observed on these organisms. Lahr et al. (1987) showed that spider numbers were reduced by ploughing but not by the use of 2,4-D herbicides.

Matida et al. (1975) examined aquatic organisms in a stream running through a mountainous area of 9.4 ha, in Shizuoka prefecture in Japan, which had been aerially sprayed with a mixture of 2,4-D and 2,4,5-T at a rate of 150 kg/ha. There was no effect on the number or species diversity of aquatic invertebrates. Caged cherry salmon and dace fingerlings showed no mortality, abnormal behaviour, or pathological change after spraying. An extensive ecological survey of an area in Florida treated with military mixtures of herbicides, including 2,4-D, revealed no major change in species diversity or population size for aquatic invertebrates, fish, a lizard, or the beach mouse (Young et al., 1975).

The weevil *Rhinocyllus conicus* is used in the biological control of musk thistle, an invasive weed. In an investigation of the practicality of combining biological with chemical control, Lee & Evans (1980) investigated the toxicity to the weevil of 2,4-D. They sprayed musk thistle with 2,4-D at a rate of 4.48 kg/ha. One week later, the terminal seed heads of the thistle were covered with cloth bags to contain the weevils. The number of dead larvae, pupae, and adults were counted and compared to the number on unsprayed, control thistle heads. No significant differences were observed.

Dwernychuk & Boag (1973) studied the effect of herbicide spraying on several species of nesting ducks (lesser scaup, gadwall, white-winged scoter, mallard, pintail, and American wigeon) in Canada. An ester of 2,4-D was applied to two islands. This application significantly reduced the areas dominated by broad-leaved plants and permitted invasion of these areas by grasses. Ducks preferred to nest amongst broad-leaved vegetation and avoided grass. As the areas of broad-leaved plants disappeared, there was an increase in nest density in those broad-leaved areas still present. Total numbers of nesting ducks declined over the 3-year study period. This decline was attributed, by the authors, entirely to the effect of the herbicide on vegetation type.

Keith et al. (1959) studied the effects on populations of the pocket gopher *(Thomomys talpoides)* of spraying weedy rangeland in Colorado with 2,4-D, as the butyl ester, at 3.4 kg/ha. Numbers of gophers were estimated using two different methods, either by trapping or by counting numbers of newly excavated mounds. Both methods showed a highly significant difference between sprayed and non-sprayed areas 1 year after spraying. The total numbers of gophers trapped in the two areas before 2,4-D application were 101 and 110, respectively. In the same two areas, 1 year after spraying, numbers were 117 (untreated area) and 15 (treated area), respectively. This represents a fall in gopher numbers of 87% in the treated area and a slight increase in numbers in the unsprayed area. Newly excavated mounds in the treated area were only 28% as numerous as in control areas. Spraying had reduced production of forbs (broad-leaved plants) from 445 kg/ha before spraying to 75 kg/ha afterwards, a reduction of 83%. Grass production had increased by 37%. The overall reduction in vegetation was 232 kg/ha or 35% on sprayed plots.

In a similar study by Tietjen et al. (1967), 2,4-D butyl ester applied at 3.4 kg/ha to identical high-altitude rangeland initially reduced forb density and gopher populations. Pocket gopher numbers were reduced by between 80% and 90%. Both forbs and gopher populations remained low in one treated area but not in a second one. The decline in gopher numbers was considered by the authors to be a result of food deficiency; the grasses available represented a marginal diet for the animals. This decline was not due either to movement of animals out of the area or to the direct toxic effects of the herbicide. Reduction in numbers was, therefore, primarily a result of reduced breeding success.

Johnson & Hansen (1969) studied the after-effects on wild mammals of treating perennial forb and shrub/grass ranges with 2,4-D either aerially or from a ground rig at rates of 2.2 or 3.4 kg/ha using a diesel-oil carrier. The density and litter size of the deer mouse *(Peromyscus maniculatus)* was little affected by the treatment, but the densities of northern pocket gophers *(Thomomys talpoides)* and least chipmunks *(Eutamias minimus)* were reduced. Montane voles *(Microtus montanus)* increased their abundance in treated perennial forb range. Gopher and vole populations returned to normal with the re-establishment of forb dominance. Density changes were considered by the authors to be primarily due to changes in availability of food for the gophers, availability of both food and cover for chipmunks, and availability of a close-canopied grass cover for the voles. There were no direct toxic effects of the herbicide.

Fagerstone et al. (1977) observed two colonies of black-tailed prairie dogs *(Cynomys ludovicianus)* in North America and the effect of spraying their feeding areas with 2,4-D. The herbicide was applied initially as the dimethylamine salt, followed by two further sprays of the butyl ester after 1 month and 1 year. All applications of 2,4-D were at 2.2 kg/ha. One colony lived in a sprayed area, initially rich in broad-leaved plants. The second colony lived in an unsprayed area poor in dicotyledonous plants and rich in grass. The effect of spraying was, therefore, to make the treated area similar to the control area with less broad-leaved herbage and less cover. Prior to treatment, the first colony preferentially ate the forbs; diet was 73% forbs and 5% grass. After spraying, they ate 9% forbs and 82% grass. The control colony ate a similar diet to the first colony post-treatment, i.e., mostly grass. Prairie dogs remained in the same area after treatment with herbicide and there was no evidence of starvation. Body weight was maintained, activity was comparable to the pre-treatment level, and reproduction was unaffected.

Spencer & Barrett (1980) monitored the population response of meadow voles *(Microtus pennsylvanicus)* to application of 2,4-D as the N-oleyl 1,3 propylenediamine salt at 567.5 g/ha. Two 0.4-ha plots were compared, one sprayed and the other untreated. The population fluctuation of the voles was monitored for 6 months, from June to December, a period spanning the breeding season. The control area reached a population peak of 116 animals on 6 November; a peak of 68 voles was reached on 9 October in the treated area. There was a skewed sex ratio in the treated area, mainly because of a reduced survival rate in females. Voles in the treated plot were protein-deficient, compared to controls.

9. EVALUATION

In evaluating the environmental hazards of 2,4-D, the following general points should be borne in mind:

(a) the chlorinated dibenzo-*p*-dioxins (CDDs) are present in 2,4-D only in trace amounts which are difficult to separate and identify;

(b) 2,4-D is rapidly degraded in the environment;

(c) the environmental effects are indirect, and are the consequence of vegetation diversity being modified; care should be taken to avoid unintentional vegetation damage; the mode of application and the formulation should be carefully selected; esters should be avoided in aquatic applications (because of toxicity to aquatic organisms);

(d) there are limited data on the effects of 2,4-D and its formulations on communities of organisms; hazard assessment is, therefore, often by extrapolation from single species studies;

(e) minor adverse effects shown in laboratory studies on terrestrial organisms have resulted from exposures far in excess of likely exposures in the field.

9.1 Aquatic Organisms

Sources of exposure of aquatic ecosystems to 2,4-D include direct application, run-off, and spray drift. Because of low adsorption rate and rapid degradation, the herbicide is not accumulated in compartments of the aquatic system.

2,4-D acid and its salts are less toxic to aquatic organisms than are the esters.

2,4-D acid and its salts are of low to moderate toxicity to aquatic organisms. However, the growth and nitrogen fixation of some cyanobacteria (blue-green algae) are inhibited, but only at concentrations above levels expected from direct application of the herbicide to water. These microorganisms are the source of most nitrogen in wet tropical soils. This inhibitory effect could be of concern when these compounds are applied to rice fields at a high dosage.

Because of the toxicity of esters, particularly the propylene glycol butyl ether ester, for early life-stages of several fish species, they should be regarded as hazardous to aquatic ecosystems.

9.2 Terrestrial Organisms

2,4-D does not persist in soil and other compartments of the terrestrial environment.

Nitrogen-fixing microorganisms appear to be particularly sensitive to 2,4-D; this might be especially important in tropical soils.

Some terrestrial invertebrates have shown adverse effects, but only at high exposure levels. Therefore, 2,4-D does not constitute a hazard to this group of organisms.

2,4-D has low acute toxicity to birds, as indicated by the LD_{50}. Most studies on birds and their eggs have been conducted at exposures exceeding those that could be expected in the field; even under these conditions, no significant adverse effects have been observed.

Under field conditions, 2,4-D does not cause direct toxic effects on animals. However, the change of species composition and structure of the vegetation, resulting from the use of this herbicide, leads to indirect effects on terrestrial ecosystems. This indirect effect would also result from the use of any herbicide or from other methods of land management in either temperate or tropical regions.

* * *

There is evidence only for minor effects on the environment arising from the use of 2,4-D, as long as the following simple recommendations are followed:

(a) amine formulations, rather than esters, should be used to control aquatic weeds;

(b) accidental spread of the herbicide to other vegetation should be avoided;

(c) the margins of agricultural land should be left untreated with herbicide to avoid even the indirect effects of the material on wildlife.

10. RECOMMENDATIONS FOR FURTHER RESEARCH

There are indications that 2,4-D affects nitrogen fixation by algae. Since this is the major source of nitrogen in tropical soils, it is recommended that this should be further investigated, particularly with reference to varying soil conditions. This should be extended to a study on the functioning of a rice-paddy at the ecosystem level.

REFERENCES

ADAMS, B.J. (1960) Effects of spraying 2,4-D amine on Coccinellid larvae. *Can. J. Zool.*, **38**: 285-288.

ALABASTER, J.S. (1969) Survival of fish in 164 herbicides, insecticides, fungicides, wetting agents and miscellaneous substances. *Int. Pest Control*, **11**: 29-35.

AUDUS, L.J. (1960) Microbiological breakdown of herbicides in soils. In: Woodford, E.K. & Sagar, G.R., ed. *Herbicides and soil*, Oxford Blackwell Scientific Publications, pp. 1-19.

AUDUS, L.J. (1964) Herbicide behaviour in the soil. In: Audus, L.J., ed. *The physiology and biochemistry of herbicides*, London, New York, Academic Press, pp. 163-206.

BAILEY, H.C. & LIU, D.H.W. (1980) *Lumbriculus variegatus*, a benthic oligochaete, as a bioassay organism. In: Eaton, J.C., Parish, P.R., & Hendricks, A.C., ed. *Aquatic toxicology*, Philadelphia, American Society for Testing and Materials, pp. 205-215 (Special Technical Publication 707).

BEDNARZ, T. (1981) The effect of 2,4-D acid on green and blue-green algae in unialgal and mixed cultures. *Acta hydrobiol.*, **23**: 173-182.

BENEDECZKY, I., BIRO, P., & SCHAFF, Z. (1984) The effect of 2,4-D-containing herbicide (Dikonirt) on the ultrastructure of carp *(Cyprinus carpio)* liver cells. *Acta biol. (Szeged)*, **30**: 107-125.

BERAN, F. & NEURURER, J. (1955) [Effect of plant protection materials on the honeybee. I. Toxicity to bees.] *Pflanzenschutzberichte*, **15**: 97-147 (in German).

BIRGE, W.J., BLACK, J.A., & BRUSER, D.M. (1979) *Toxicity of organic chemicals to embryo-larval stages of fish.* Washington, DC, US Environmental Protection Agency (EPA 560-1179-007).

BIRO, P. (1979) Acute effects of the sodium salt of 2,4-D on the early developmental stages of bleak, *Alburnus alburnus*. *J. Fish Biol.*, **14**: 101-109.

BJORKLUND, N-E. & ERNE, K. (1966) Toxicological studies of phenoxyacetic herbicides in animals. *Acta vet. Scand.*, **7**: 364-390.

BJORN, M.K. & NORTHEN, H.T. (1948) Effects of 2,4-dichlorophenoxyacetic acid on chicks. *Science*, **108**: 479-480.

BUNTING, D.L. & ROBERTSON, E.B. (1975) *Lethal and sublethal effects of herbicides on zooplankton species,* University of Tennessee, Water Resources Research Center, 34 pp. (Res. Report No. 43, NTIS Report No. PB-241 337).

BUTLER, G.L., DEASON, T.R., & O'KELLEY, J.C. (1975a) Loss of five pesticides from cultures of twenty-one planktonic algae. *Bull. environ. Contam. Toxicol.,* **13**: 149-152.

BUTLER, G.L., DEASON, T.R., & O'KELLEY, J.C. (1975b) The effect of atrazine, 2,4-D, methoxychlor, carbaryl and diazinon on the growth of planktonic algae. *Br. phycol. J.,* **10**: 371-376.

BUTLER, P.A. (1963) *A review of Fish and Wildlife Service investigations during 1961 and 1962,* Washington, DC, US Department of the Interior, Fish and Wildlife Service, pp. 11-25 (Circular No. 167).

BUTLER, P.A. (1965) Effects of herbicides on estuarine fauna. *Proc. South. Weed Conf.,* **18**: 576-580.

CALDWELL, R.S. (1977) *Biological effects of pesticides on the Dungeness crab,* Gulf Breeze, Florida, US Environmental Protection Agency (EPA-600/3-77-131).

CALDWELL, R.S., BUCHANAN, D.V., ARMSTRONG, D.A., MALLON, M.H., & MILLEMANN, R.E. (1979) Toxicity of the herbicides 2,4-D, DEF, propanil and trifluarin to the Dungeness crab, *Cancer magister. Arch. environ. Contam. Toxicol.,* **8**: 383-386.

CHAI, I.K. & CHUNG, Y.S. (1975) [Physiological effects of 2,4-dichlorophenoxyacetic acid (2,4-D) on *Chlorella ellipsoidea.] Misaengmul Hakhoe Chi.,* **13**: 101-108 (in Korean).

CHAMBERS, H., DZIUK, L.J., & WATKINS, J. (1977) Hydrolytic activation and detoxification of 2,4-dichlorophenoxyacetic acid esters in mosquito fish. *Pestic. Biochem. Physiol.,* **7**: 297-300.

CHEAH, M.L., AVAULT, J.W., & GRAVES, J.B. (1980) Acute toxicity of selected rice pesticides to crayfish *Procambarus clarkii. Prog. Fish Cult.,* **42**: 169-172.

CHOLAKIS, J.M., MCKEE, M.J., WONG, L.C.K., & GILE, J.D. (1982) Acute and subacute toxicity of pesticides in microtine rodents. In: Lamb, D.W. & Kenaga, E.E., ed. *Avaian and Mammalian Wildlife Toxicology: 2nd Conference,* Philadelphia, American Society for Testing and Materials, pp. 143-154 (Special Technical Publication 757).

COAKLEY, J.E., CAMPBELL, J.E., & MCFARREN, E.F. (1964) Determination of butoxyethanol ester of 2,4-dichlorophenoxyacetic acid in shellfish and fish. *J. agric. food Chem.,* **12**: 262-265.

COOKE, A.S. (1972) The effects of DDT, dieldrin and 2,4-D on amphibian spawn and tadpoles. *Environ. Pollut.*, **3**: 51-68.

COPE, O.B., WOOD, E.M., & WALLEN, G.H. (1970) Some chronic effects of 2,4-D on the bluegill *(Lepomis macrochirus). Trans. Am. Fish. Soc.*, **99**: 1-12.

CULLIMORE, D.R. & MCCANN, A.E. (1977) Influence of four herbicides on the algal flora of a prairie soil. *Plant Soil*, **46**: 499-510.

DAS, B. & SINGH, P.K. (1977) The effect of 2,4-dichlorophenoxyacetic acid on growth and nitrogen-fixation of blue-green alga *Anabaenopsis raciborskii. Arch. environ. Contam. Toxicol.*, **5**: 437-445.

DAVIS, J.T. & HARDCASTLE, W.S. (1959) Biological assay of herbicides for fish toxicity. *Weeds*, **7**: 397-404.

DAVIS, P.W., FRIEDHOFF, J.M., & WEDERMEYER, G.A. (1972) Organochlorine insecticide, herbicide and polychlorinated biphenyl (PCB) inhibition of NaK-ATPase in rainbow trout. *Bull. environ. Contam. Toxicol.*, **8**: 69-72.

DESHMUKH, V.A. & SHRIKHANDE, J.G. (1974) Effect of some herbicides on soil microflora. *J. Indian Soc. Soil Sci.*, **22**: 300-303.

DESHMUKH, V.A. & SHRIKHANDE, J.G. (1975) Effect of some herbicides on Azotobacter and non-symbiotic nitrogen fixation in the soil. *Indian J. Agric. Chem.*, **8**: 95-98.

DEWITT, J.B., STICKEL, W.H., & SPRINGER, P.F. (1963) *Wildlife studies, Patuxent Wildlife Research Center. Pesticide-wildlife studies. A review of Fish and Wildlife Service investigations during 1961 and 1962.* Washington, DC, US Department of the Interior, Fish and Wildlife Service, pp. 74-96 (Circular No. 167).

DODSON, J.J. & MAYFIELD, C.I. (1979) The dynamics and behavioural toxicology of Aqua-Kleen (2,4-D butoxyethanol ester) as revealed by the modification of rheotropism in rainbow trout. *Trans. Am. Fish. Soc.*, **108**: 632-640.

DUNACHIE, J.F. & FLETCHER, W.W. (1967) Effect of some herbicides on the hatching rate of hen's eggs. *Nature (Lond.)*, **215**: 1406-1407.

DWERNYCHUK, L.W. & BOAG, D.A. (1973) Effect of herbicide-induced changes in vegetation on nesting ducks. *Can. field Nat.*, **87**: 155-165.

ELDER, J.H., LEMBI, C.A., & MORRE, D.J. (1970) Toxicity of 2,4-D and picloram to fresh and saltwater algae. *Proc. North-cent. Weed Control Conf.*, **25**: 96-98.

ELIASSON, L. (1973) Translocation and persistence of 2,4-D in *Populus tremula* L. *Weed Res.*, **13**: 140-147.

ERNE, K. & BJORKLUND, N-E. (1970) Nephrotoxic effects of phenoxy acetic herbicides. In: *Proceedings of the 7th International Congress on Plant Protection, Paris, France*, pp. 768-769.

EVERTS, J.W., BEURSKENS, J.E.M., BOUWHUIS, J.S., BUYSE, A.D., HENGEVELD, R., WOUTERS, L., & KOEMAN, J.H. (1986) Terrestrial arthropods as indicators for side-effects caused by insecticides in arable farm systems in the Netherlands. In: *Proceedings of the 1st International TNO Conference on Contaminated Soil, Utrecht, The Netherlands, November 1985*, Dordrecht, Boston, Martinus Nijhoff, pp. 423-425.

FAGERSTONE, K.A., TIETJEN, H.P., & LAVOIE, G.K. (1977) Effects of range treatment with 2,4-D on prairie dog diet. *J. Range Manage.*, **30**: 57-60.

FINLAYSON, B.J. & VERRUE, K.M. (1985) Toxicities of butoxyethanol ester and propylene glycol butyl ether ester formulations of 2,4-dichlorophenoxy acetic acid (2,4-D) to juvenile salmonids. *Arch. environ. Contam. Toxicol.*, **14**: 153-160.

FLINT, G.W., ALEXANDER, J.J., & FUNDERBURK, O.P. (1968) Vapor pressures of low-volatile esters of 2,4-D. *Weed Sci.*, **16**: 541-544.

FOLMAR, L.C. (1976) Overt avoidance reaction of rainbow trout fry to nine herbicides. *Bull. environ. Contam. Toxicol.*, **15**: 509-514.

FOLMAR, L.C. (1978) Avoidance chamber responses of mayfly nymphs exposed to eight herbicides. *Bull. environ. Contam. Toxicol.*, **19**: 312-318.

FOURNIER, J.C. (1980) Enumeration of the soil micro-organisms able to degrade 2,4-D by metabolism or co-metabolism. *Chemosphere*, **9**: 169-174.

FREITAG, D., GEYER, H., KRAUS, A., VISWANATHAN, R., KOTZIAS, D., ATTAR, A., KLEIN, W., & KORTE, F. (1982) Ecotoxicological profile analysis. VII, Screening chemicals for their environmental behaviour by comparative evaluation. *Ecotoxicol. environ. Saf.*, **6**: 60-81.

GALL, A. & DOGGER, J.R. (1967) Effect of 2,4-D on the wheat stem sawfly. *J. econ. Entomol.*, **60**: 75-77.

GANGAWANE, L.V., SALER, R.S., & KULKARNI, L. (1980) Effect of pesticides on growth and heterocyst formation in *Nostoc sp. Marathwada Univ. J. Sci.*, **19B**: 3-6.

GEORGE, J.P., HINGORANI, H.G., & RAO, K.S. (1982) Herbicide toxicity to fish-food organisms. *Environ. Pollut.*, **28**: 183-188.

GILE, J.D. (1983) 2,4-D - its distribution and effects in a ryegrass ecosystem. *J. environ. Qual.*, **12**: 406-412.

GLYNN, P.W., HOWARD, L.S., CORCORAN, E., & FREAY, A.D. (1984) The occurrence and toxicity of herbicides in reef-building corals. *Mar. Pollut. Bull.*, **15**: 370-374.

GRAHAM-BRYCE, I.J. (1972) Herbicide movement and availability in soils. *Proc. Br. Weed Control Conf.*, **11**: 1193-1202.

GROLLEAU, G., DE LAVAUR, E., & SIOU, G. (1974) Effets du 2,4-D sur la reproduction des cailles et des perdrix, après application du produit par pulvérisation sur les oeufs. *Ann. Zool. Ecol. anim.*, **6**: 313-331.

GUARINO, A.M., JAMES, M.O., & BEND, J.R. (1977) Fate and distribution of the herbicides 2,4-dichlorophenoxyacetic acid (2,4-D) and 2,4,5-trichlorophenoxyacetic acid (2,4,5-T) in the dogfish shark. *Xenobiotica*, **7**: 623-631.

GYRD-HANSEN, N. & DALGAARD-MIKKELSEN, S. (1974) The effect of phenoxy-herbicides on the hatchability of eggs and the viability of the chicks. *Acta pharmacol. toxicol.*, **35**: 300-308.

HAEGELE, M.A. & TUCKER, R.K. (1974) Effects of 15 common environmental pollutants on eggshell thickness in mallards and coturnix. *Bull. environ. Contam. Toxicol.*, **11**: 98-102.

HAGENMAIER, H. (1986) Determination of 2,3,7,8-tetrachlorodibenzo-p-dioxin in commercial chlorophenols and related products. *Fresenius Z. anal. Chem.*, **325**: 603-606.

HANSEN, D.J., SCHIMMEL, S.C., & KELTNER, J.M. (1973) Avoidance of pesticides by grass shrimp *(Palaemonetes pugio)*. *Bull. environ. Contam. Toxicol.*, **9**: 129-133.

HAQUE, A. & EBING, W. (1983) Uptake, accumulation, and elimination of HCB and 2,4-D by the terrestrial slug, *Deroceras reticulatum* (Muller). *Bull. environ. Contam. Toxicol.*, **31**: 727-733.

HASHIMOTO & NISHIUCHI (1978) quoted by Hidaka et al. (1984).

HAWXBY, K., TUBEA, B., OWNBY, J., & BASLER, E. (1977) Effects of various classes of herbicides on four species of algae. *Pestic. Biochem. Physiol.*, **7**: 203-209.

HIDAKA, H., HATTANDA, M., & TATSUKAWA, R. (1984) [Avoidance of pesticides with medakas *(Oryzias latipes).]* *Nippon Nogeikagaku Kaishi*, **58**: 145-151 (in Japanese).

HILBIG, V., LUCAS, K., & SEBEK, V. (1976a) [Studies on the susceptibility of eggs of gallinaceous birds to 2,4,5-T and 2,4-D (First report).] *Anz. schädlingskd. Pflanzenschutz Umweltschutz*, **49**: 21-25 (in German).

HILBIG, V., LUCAS, K., & SEBEK, V. (1976b) [Studies on the susceptibility of eggs of gallinaceous birds to 2,4,5-T and 2,4-D (Second report).] *Anz. schädlingskd. Pflanzenschutz Umweltschutz*, **49**: 65-68 (in German).

HILL, E.F., HEATH, R.G., SPANN, J.W., & WILLIAMS, J.D. (1975) *Lethal dietary toxicities of environmental pollutants to birds*, Washington, DC, US Department of the Interior, Fish Wildlife Service, 61 pp. (Special Science Report, Wildlife No. 191).

HILTIBRAN, R.C. (1967) Effects of some herbicides on fertilized fish eggs and fry. *Trans. Am. Fish. Soc.*, **96**: 414-416.

HILTIBRAN, R.C. (1975) Oxygen and phosphate metabolism of bluegill liver mitochondria in the presence of some pesticides. *Environ. Qual. Saf. (Suppl)*, **3**: 549-553.

HOEPPEL, R.E. & WESTERDAHL, H.E. (1983) Dissipation of 2,4-D DMA and BEE from water, mud and fish at Lake Seminole, Georgia. *Water Resour. Bull.*, **19**: 197-204.

HOFFMAN, D.J. & ALBERS, P.H. (1984) Evaluation of potential embryotoxicity and teratogenicity of 42 herbicides, insecticides and petroleum contaminants to mallard eggs. *Arch. environ. Contam. Toxicol.*, **13**: 15-27.

HUBER, S.J., POSCHENREIDER, G., & WALLNOFER, P.R. (1980) [Effect of pesticides and the corresponding metabolites on growth and respiration of some soil microorganisms.] *Z. Pflanzenkr. Pflanzenschutz*, **87**: 533-545 (in German).

HUDSON, R.H., TUCKER, R.K., & HAEGELE, M.A. (1984) *Handbook of toxicity of pesticides to wildlife, 2nd edition*, Washington, DC, US Department of the Interior, Fish and Wildlife Service, pp. 90 (Resource Publication 153).

HUGHES, J.S. & DAVIS, J.T. (1963) Variations in toxicity to bluegill sunfish of phenoxy herbicides. *Weeds*, **11**: 50-53.

JAMES, M.O. (1979) Disposition of 2,4-D, 2,4,5-T, DDA and phenylacetic acid in the spiny lobster *(Panulirus argus)*. *Pharmacologist*, **21**: 251.

JANA, B.B. & DE, U.K. (1982) Stimulatory effect of herbicide 2,4-D on the heterotrophic microbial community in the water of three fish ponds. *Acta microbiol. Acad. Sci. Hung.*, **29**: 77-82.

JENSEN, D.A. & SCHALL, E.D. (1966) Determination of vapor pressures of some phenoxyacetic herbicides by gas-liquid chromatography. *J. agric. food Chem.*, **14**: 123-125.

JOHNSON, C.R. (1976) Herbicide toxicities in some Australian anurans and the effect of subacute dosages on temperature tolerance. *Zool. J. Linnean Soc.*, **59**: 79-83.

JOHNSON, C.R. (1978) Herbicide toxicities in the mosquito fish *Gambusia affinis*. *Proc. R. Soc. Queensland.*, **89**: 25-27.

JOHNSON, D.R. & HANSEN, R.M. (1969) Effects of range treatment with 2,4-D on rodent populations. *J. wildl. Manage.*, **33**: 125-132.

JONES, G.D.E. & CONNELL, J.U. (1954) Studies of the toxicity to worker honey-bees *(Apis mellifera* L.) of certain chemicals used in plant protection. *Ann. appl. Biol.*, **41**: 271-279.

KAPOOR, K. & SHARMA V.K. (1980) Effect of certain herbicides on survival, growth and nitrogen fixation of blue-green alga *Anabaena doliolum Bharadwaja*. *Z. allg. Mikrobiol.*, **20**: 465-469.

KEITH, J.O., HANSEN, R.M., & WARD, A.L. (1959) Effect of 2,4-D on abundance and foods of pocket gophers. *J. wildl. Manage.*, **23**: 137-145.

KOPISCHKE, E.D. (1972) Effect of 2,4-D and diesel fuel on egg hatchability. *J. wildl. Manage.*, **36**: 1353-1356.

KOSCHIER, F.J. & PRITCHARD, J.B. (1980) Renal handling of 2,4-D by dogfish shark *(Squalus acanthias)*. *Xenobiotica*, **10**: 1-6.

LAHR, J., GOEDEJOHAN, N., & EVERTS, J.W. (1987) Side effects of pesticides on terrestrial arthropods. I. Long-term effects, geographical variability and effects of cultural practices. In: *Proceedings of the Symposium "25 Years After Silent Spring", November 1987*, Amsterdam, Toxicological Society.

LE VAN TO (1984) [Degradation of defoliants 2,4-D and 2,4,5-T by selected soil microorganisms.] *Acta agrar. Silvestria, Ser. agrar.*, **23**: 225-234 (in Polish).

LEE, R.D. & EVANS, J.O. (1980) The influence of selected herbicides on the development of *Rhinocyllus conicus*, an insect used in biocontrol of musk thistle. *Proc. West. Soc. Weed Sci.*, **33**: 104-110.

LEMBI, C.A. & COLERIDGE, S.E. (1975) *Selective toxicity of detergents and herbicides to phytoplankton*, West Lafayette, Indiana, Purdue University, Water Resources Research Center, 71 pp. (Technical Report No. 71).

LEWIS, D.L., HODSON, R.E., & FREEMAN, L.F. (1984) Effects of microbial community interactions on transformation rates of xenobiotic chemicals. *Appl. environ. Microbiol.*, **48**: 561-565.

LHOSTE, J. & ROTH, P. (1946) Sur l'action des solutions aqueuses de 2,4-dichlorophénoxyacétate de sodium sur l'évolution des oeufs de *Rana temporaria* L. *Soc. Biol.*, **140**: 272-273.

LIEBERMAN, M.T. & ALEXANDER, M. (1981) Effects of pesticides on decomposition of organic matter and nitrification in sewage. *Bull. environ. Contam. Toxicol.*, **26**: 554-560.

LINDEN, E., BENGTSSON, B-E., SVANBERG, O., & SUNDSTROM, G. (1979) The acute toxicity of 78 chemicals and pesticide formulations against two brackish water organisms, the bleak *(Alburnus alburnus)* and the harpacticoid *Nitocra spinipes*. *Chemosphere*, **8**: 843-851.

LI-TSE OU (1984) 2,4-D degradation and 2,4-D degrading microorganisms in soils. *Soil Sci.*, **137**: 100-107.

LIU, H.W. & LEE, M. (1975) *Toxicity of selected pesticides to the bay mussel Mytilus edulis*, Washington, DC, US Department of Commerce, National Technical Information Service (NTIS Report No. PB-243 221).

LOOS, M.A. (1969) Phenoxyalkanoic acids. In: Kearney, P.C. & Kaufman, D.D., ed. *Degradation of herbicides*, New York, Marcel Dekker, Inc., pp. 1-49.

LUNDHOLM, C.E. & MATHSON, K. (1983) The inhibitory effect of some chlorinated hydrocarbon pesticides on the ATP-dependent Ca^{2+} binding of the particulate fraction of the eggshell gland mucosa cells. *Acta pharmacol. toxicol.*, **52**: 390-394.

LUTZ, H. & LUTZ-OSTERTAG, Y. (1972) The action of different pesticides on the development of bird embryos. *Adv. exp. Med. Biol.*, **27**: 127-150.

LUTZ, H. & LUTZ-OSTERTAG, Y. (1973) Pesticides, tératogénèse et survie chez les oiseaux. *Arch. Anat. Histol. Embryol.*, **56**: 65-78.

LUTZ-OSTERTAG, Y. & LUTZ, H. (1970) Action néfaste de l'herbicide 2,4-D sur le développement embryonnaire et la fécondité du gibier à plumes. *C. R. Acad. Sci. Paris, Ser. D*, **271**: 2418-2421.

MATIDA, Y., FURUTA, Y., KUMADA, H., TANAKA, H., YOKOTE, M., & KIMURA, S. (1975) Effects of some herbicides applied in the forest to the freshwater fishes and other aquatic organisms - I. Survey on the effects of aerially applied sodium chlorate and a mixture of 2,4-D and 2,4,5-T on the stream community. *Bull. freshwater Fish Res. Lab. (Tokyo)*, **25**: 41-53.

MAYER, F.L. (1987) *Acute toxicity handbook of chemicals to estuarine organisms*, Washington, DC, US Department of Commerce, National Technical Information Service, 274 pp. (NTIS Report No. PB-87 188686).

MAYER, F.L. & ELLERSIECK, M.R. (1986) *Manual of acute toxicity: interpretation and data base for 410 chemicals and 66 species of freshwater animals*, Washington, DC, US Department of the Interior, Fish and Wildlife Service, 506 pp. (Resource Publication 160).

MEEHAN, W.R., NORRIS, L.A., & SEARS, H.S. (1974) Toxicity of various formulations of 2,4-D to salmonids in southeast Alaska. *J. Fish Res. Board Can.*, **31**: 480-485.

METCALF, R.L. & SANBORN, J.R. (1975) Pesticides and environmental quality in Illinois. *Illinois nat. Hist. Surv. Bull.*, **31**: 381-435.

MORTON, H.L., MOFFETT, J.O., & MACDONALD, R.H. (1972) Toxicity of herbicides to newly emerged honey bees. *Environ. Entomol.*, **1**: 102-104.

MOUBASHER, A.H., EL-HISSY, F.T., & ABDEL-KADER, M.I.A. (1981) Selective effects of three herbicides on the fungus flora of Egyptian soil. *Folia microbiol.*, **26**: 37-44.

MOUNT, D.I. & STEPHAN, C.E. (1967) A method for establishing acceptable toxicant limits for fish - malathion and the butoxyethanol ester of 2,4-D. *Trans. Am. Fish. Soc.*, **96**: 185-193.

MUKHOPADHYAY, S.K. (1980) Effects of herbicides and insecticides alone and their combinations on soil microflora. *Indian J. Weed Sci.*, **12**: 53-60.

MULLER, G. (1971) [Laboratory studies on effects of herbicides on Carabidae.] *Arch. Pflanzenschutz*, **7**: 351-364 (in German).

NAGUIB, K., YOUNIS, A.E., & EL-ESSAWY, M.A. (1980) The effects of 2,4-dichlorophenoxyacetic acid on respiration, nitrogen and carbohydrate metabolism of *Aspergillus terreus* Thom. *Zbl. Bakteriol. II. Abt.*, **135**: 252-259.

NESBITT, H.J. & WATSON, J.R. (1980) Degradation of the herbicide 2,4-D in river water. II. The role of suspended sediment, nutrients and water temperature. *Water Res.*, **14**: 1689-1694.

NORRIS, L.A. & GREINER, D. (1967) The degradation of 2,4-D in forest litter. *Bull. environ. Contam. Toxicol.*, **2**: 65-74.

O'CONNOR, G.A., FAIRBANKS, B.C., & DOYLE, E.A. (1981) Effects of sewage sludge on phenoxy herbicide adsorption and degradation in soils. *J. environ. Qual.*, **10**: 510-515.

OKA, I.N. & PIMENTAL, D. (1976) Herbicide (2,4-D) increases insect and pathogen pests on corn. *Science*, **193**: 239-240.

OMKAR & SHUKLA, G.S. (1984) Toxicity of the herbicide 2,4-D-Na to two species of freshwater prawn of the genus *Macrobranchium*. *Acta hydrochim. hydrobiol.*, **12**: 285-289.

PACHPANDE, R.R. & DAVID, S.B. (1980) Effect of selected herbicides on the growth of soil alga *Chlorococcum infusionum meneghini* (Schrank). *Phykos*, **19**: 138-141.

PALMER, J.S. (1972) *Toxicity of 45 organic herbicides to cattle, sheep, and chickens*, Washington, DC, US Department of Agriculture, Agricultural Research Service, 41 pp. (Production Research Report No. 137).

PALMER, J.S. & RADELEFF, R.D. (1969) *The toxicity of some organic herbicides to cattle, sheep, and chickens*. Washington, DC, US Department of Agriculture, Agricultural Research Service, 26 pp. (Production Research Report No. 106).

PLUMB, T.R., NORRIS, L.A., & MONTGOMERY, M.L. (1977) Persistence of 2,4-D and 2,4,5-T in chaparral soil and vegetation. *Bull. environ. Contam. Toxicol.*, **17**: 1-8.

PONS, R. & PUSSARD, M. (1980) Action des herbicides sur les amibes libres (Rhizopoda, Protozoa). Etude préliminaire. *Acta oecol. oegol. appl.*, **1**: 15-20.

POORMAN, A.E. (1973) Effects of pesticides on Euglena gracilis. I. Growth studies. *Bull. environ. Contam. Toxicol.*, **10**: 25-28.

PRESCOT, L.M. & OLSON, D.L. (1972) Effects of pesticides on the soil amoeba, *Acanthamoeba castellanii*. *Proc. South Dakota Acad. Sci.*, **51**: 136-141.

PRESING, M. (1981) On the effects of Dikonirt (sodium salt of 2,4-dichlorophenoxyacetic acid) on the mortality and reproduction of *Daphnia magna*. *Hydrobiologia*, **83**: 511-516.

PRITCHARD, J.B. & JAMES, M.O. (1979) Determinants of the renal handling of 2,4-dichlorophenoxyacetic acid by winter flounder. *J. Pharmacol. exp. Ther.*, **208**: 280-286.

RAATIKAINEN, M., SILTANEN, H., ROSENBERG, C., RAATIKAINEN, T., & MUKULA, J. (1979) Herbicide residues in cowberries, bilberries and lichens in controlled ground spraying experiments on woodland. *Ann. agric. Fenniae*, **18**: 112-116.

RADOSEVICH, S.R. & WINTERLIN, W.L. (1977) Persistence of 2,4-D and 2,4,5-T in chaparral vegetation and soil. *Weed Sci.*, **25**: 423-425.

RAND, G.M. & BARTHALMUS, G.T. (1980) *Use of an unsignalled avoidance technique to evaluate the effects of the herbicide 2,4-dichlorophenoxyacetic acid on goldfish*, Philadelphia, American Society for Testing and Materials, pp. 341-353 (Special Technical Publication 707).

RAPOPORT, E.H. & CANGIOLI, G. (1963) Herbicides and the soil fauna. *Pedobiologia*, **2**: 235-238.

REHWOLDT, R.E., KELLEY, E., & MAHONEY, M. (1977) Investigations into the acute toxicity and some chronic effects of selected herbicides and pesticides on several fresh water fish species. *Bull. environ. Contam. Toxicol.*, **18**: 361-365.

RIVIERE, J-L. (1976) Les effets secondaires des pesticides agricoles. II. Action de quelques herbicides sur la croissance et la fécondité de *Blattella germanica* (L.). *Ann. Zool. Ecol. anim.*, **8**: 543-550.

ROBERTS, B.L. & DOROUGH, H.W. (1984) Relative toxicities of chemicals to the earthworm *Eisenia foetida*. *Environ. Toxicol. Chem.*, **3**: 67-78.

ROBERTSON, E.B. (1975) *The acute toxicity of four herbicides to 0 - 4 hour nauplii of the copepod Cyclops vernalis Fisher*, Washington, DC, US Department of Commerce, National Technical Information Service, 70 pp. (M.S. Thesis, University of Tennessee) (NTIS Report No. PB-269 495).

RODGERS, C.A. & STALLING, D.L. (1972) Dynamics of an ester of 2,4-D in organs of three fish species. *Weed Sci.*, **20**: 101-105.

ROGERS, J.G., CAGAN, R.H., & KARE, M.R. (1974) Percutaneous absorption of several chemicals, some pesticides included, in the red-winged blackbird. *Environ. Physiol. Biochem.*, **4**: 104-111.

ROSENBERG, A. & ALEXANDER, M. (1980) 2,4,5-trichlorophenoxy-acetic acid (2,4,5-T) decomposition in tropical soil and its co-metabolism by bacteria *in vitro*. *J. agric. food Chem.*, **28**: 705-709.

ROWE, V.K. & HYMAS, T.A. (1954) Summary of toxicological information on 2,4-D and 2,4,5-T type herbicides and an evaluation of the hazards to livestock associated with their use. *Am. J. vet. Res.*, **15**: 622-629.

RUGGIERO, P. & RADOGNA, V.M. (1985) Inhibition of soil humus-laccase complexes by some phenoxyacetic and *s*- triazine herbicides. *Soil Biol. Biochem.*, **17**: 309-312.

SANDERS, H.O. (1969) *Toxicity of pesticides to the crustacean, Gammers lacustris*. Washington, DC, US Department of the Interior, Fish and Wildlife Service, Bureau of Sport, Fisheries and Wildlife, 18 pp. (Technical Paper No. 25).

SANDERS, H.O. (1970a) Toxicities of some herbicides to six species of freshwater crustaceans. *J. Water Pollut. Control Fed.*, **42**: 1544-1550.

SANDERS, H.O. (1970b) Pesticide toxicities to tadpoles of the western chorus frog *Pseudacris triseriata* and Fowler's toad *Bufo woodhousii fowleri*. *Copeia*, **1970**(2): 246-251.

SANDERS, H.O. & COPE, O.B. (1966) Toxicities of several pesticides to two species of cladocerans. *Trans. Am. Fish. Soc.*, **95**: 165-169.

SANDERS, H.O. & COPE, O.B. (1968) The relative toxicities of several pesticides to naiads of three species of stoneflies. *Limnol. Oceanogr.*, **13**: 112-117.

SANDMANN, E.R.I.C. & LOOS, M.A. (1984) Enumeration of 2,4-D-degrading microorganisms in soils and crop plant rhizospheres using indicator media; high populations associated with sugarcane *(Saccharum officinarum)*. *Chemosphere*, **13**: 1073-1084.

SARMA, Y.S.R.K. & TRIPATHI, S.N. (1980) Effects of some chemicals (2,4-dichlorophenoxyacetic acid, coumarin and acenaphthene) on the green alga, *Oedogonium acmandrium Elfving*. *Phykos*, **19**: 142-152.

SATTAR, M.A. & PAASIVIRTA, J. (1980) Fate of chlorophenoxyacetic acids in acid soil. *Chemosphere*, **9**: 745-752.

SCHRODER, H. & PILZ, E. (1983) [Effect of herbicides on nitrification in different soils.] *Zbl. Mikrobiol.*, **138**: 223-234 (in German).

SCHRODER, I., MEYER, M., & MUCKE, D. (1970) [Effects of the herbicides 2,4-D, aminotriazole, atrazine, chlorpropham and chlorfurenol on nucleic-acid biosynthesis in the ascomycete *Neurospora crassa.]* *Weed Res.*, **10**: 172-177 (in German).

SCHULTZ, D.P. (1973) Dynamics of a salt of (2,4-dichlorophenoxy)-acetic acid in fish, water, and hydrosol. *J. agric. food Chem.*, **21**: 186-192.

SCHULTZ, D.P. & HARMAN, P.D. (1974) Residues of 2,4-D in pond waters, mud, and fish, 1971. *Pestic. monit. J.*, **8**: 173-179.

SCHULTZ, D.P. & WHITNEY, E.W. (1974) Monitoring 2,4-D residues at Loxahatchee National Wildlife Refuge. *Pestic. monit. J.*, **7**: 146-152.

SEIBERT, K., FUHR, F., & MITTELSTAEDT, W. (1982) Experiments on the influence of roots and soils on 2,4-D degradation. *Landwirtsch. Forsch.*, **35**: 5-13.

SHCHERBAKOV, Y.A. & POLUBOYARINOVA, I.V. (1973) The persistence of 2,4-D butyl ester in water and its accumulation in aquatic life. In: Andrushaitis, G.P., ed. *Experimental water toxicology*, Charlotte, North Carolina, Army Foreign Service and Technology Center, pp. 19-21 (NTIS Report No. AD-760 400).

SHUKLA, G.S. & OMKAR (1983) Toxicity of 2,4-D-Na salt to fresh water prawn, *Macrobrachium lamarrei* (M. Edwards). *Comp. Physiol. Ecol.*, **8**: 282-284.

SIGMON, C. (1979) Oxygen consumption in Lepomis macrochirus exposed to 2,4-D or 2,4,5-T. *Bull. environ. Contam. Toxicol.*, **21**: 826-830.

SIKKA, M.C., APPLETON, H.T., & GANGSTAD, E.O. (1977) Uptake and metabolism of dimethylamine salt of 2,4-D acid by fish. *J. agric. food Chem.*, **25**: 1030-1033.

SILTANEN, H., ROSENBERG, C., RAATIKAINEN, M., & RAATIKAINEN, T. (1981) Triclopyr, glycophosphate, and phenoxyherbicide residues in cowberries, bilberries and lichen. *Bull. environ. Contam. Toxicol.*, **27**: 731-737.

SINGH, P.K. (1974) Algicidal effect of 2,4-dichlorophenoxy acetic acid on blue-green alga *Cylindrospermum sp. Arch. Microbiol.*, **97**: 69-72.

SKOKOVA, N.N. (1975) [Effect of the butyl ester of 2,4-D on the reproductive function of male bank voles.] *Nauchn. Osn. Okhr. Prir.*, **3**: 172-178 (in Russian).

SMITH, G.E. & ISOM, B.G. (1967) Investigation of effects of large-scale applications of 2,4-D on aquatic fauna and water quality. *Pestic. monit. J.*, **1**: 16-21.

SOLOMON, K.E., DAHLGREN, R.B., & LINDER, R.L. (1973) Abnormal embryos in eggs of pheasants given 2,4-D or PCB. *Proc. South Dakota Acad. Sci.*, **52**: 95-99.

SOMERS, J.D., MORAN, E.T., & REINHART, B.S. (1974) Effect of external application of pesticides to the fertile egg on hatching success and early chick performance. 3. Consequences of combining 2,4-D with picloram and extremes in contamination. *Bull. environ. Contam. Toxicol.*, **11**: 511-516.

SOMERS, J.D., MORAN, E.T., & REINHART, B.S. (1978a) Hatching success and early performance of chicks from eggs sprayed with 2,4-D, 2,4,5-T and picloram at various stages of embryonic development. *Bull. environ. Contam. Toxicol.*, **20**: 289-293.

SOMERS, J.D., MORAN, E.T., & REINHART, B.S. (1978b) Reproductive success of hens and cockerels originating from eggs sprayed with 2,4-D, 2,4,5-T and picloram followed by early performance of their progeny after a comparable *in ovo* exposure. *Bull. environ. Contam. Toxicol.*, **20**: 111-119.

SPAIN, J.C. & VAN VELD, P.A. (1983) Adaptation of natural microbial communities to degradation of xenobiotic compounds: Effects of concentration, exposure time, inoculum and chemical structure. *Appl. environ. Microbiol.*, **45**: 428-435.

SPENCER, S.R. & BARRETT, G.W. (1980) Meadow vole population responses to vegetational changes resulting from 2,4-D application. *Am. Midl. Nat.*, **103**: 32-46.

SPITTLER, H. (1976) [Experiments on the influence of herbicides on the embryo development of *Phasianus colchicus* var. and *Coturnix coturnix japonica.]* *Z. Jagdwiss.*, **22**: 197-201 (in German).

STALLING, D.L. & HUCKINS, J.N. (1978) Metabolism of 2,4-dichlorophenoxyacetic acid (2,4-D) in bluegills and water. *J. agric. food Chem.*, **26**: 447-452.

STATHAM, C.N. & LECH, J.J. (1975) Potentiation of the acute toxicity of several pesticides and herbicides in trout by carbaryl. *Toxicol. appl. Pharmacol.*, **34**: 83-87.

TAG EL-DIN, A., ABBAS, M.M., ALY, H.A., TANTAWY, G., & ASKAR, A. (1981) Acute toxicities to *Mugil cephalus* fry caused by some herbicides and new pyrethroids. *Meded. Fac. Landbouwwet. Rijksuniv. Gent*, **46**: 387-391.

THOMPSON, D.G., STEPHENSON, G.R., SOLOMON, K.R., & SKEPASTS, A.V. (1984) Persistence of (2,4-dichlorophenoxy) acetic acid and 2-(2,4-dichlorophenoxy) propionic acid in agricultural and forest soils of northern and southern Ontario. *J. agric. food Chem.*, **32**: 578-581.

TIETJEN, H.P., HALVORSON, C.H., HEGDAL, P.L., & JOHNSON, A.M. (1967) 2,4-D herbicide, vegetation, and pocket gopher relationships Black Mesa, Colorado. *Ecology*, **48**: 634-643.

TIWARI, D.N., PANDEY, A.K., & MISRA, A.K. (1984) Toxicity of 2,4-dichlorophenoxy acetic acid on growth and nitrogen fixation of blue-green alga *Anabaena cylindrica*. *Pesticides (Bombay)*, **18**: 16-18.

TOOBY, T.E., LUCEY, J., & STOTT, B. (1980) The tolerance of grass carp *Ctenopharyngodon idella* Val., to aquatic herbicides. *J. Fish Biol.*, **16**: 591-597.

TORSTENSSON, N.T.L. (1975) Degradation of 2,4-D and MCPA in soils of low pH. *Environ. Qual. Saf. (Suppl.)*, **3**: 262-265.

TOURE, I.M. & STENZ, E. (1977) [The action of selected herbicides on bacteriophages and *Escherichia coli.]* *Zbl. Bakteriol. Parasitol. Infektionskr.* **132**: 163-177 (in German).

TREVORS, J.T. & STARODUB, M.E. (1983) Effect of 2,4-D on electron transport system (ETS) activity and respiration in soil. *Bull. environ. Contam. Toxicol.*, **31**: 595-598.

TRUMBLE, J.T. & KOK, L.T. (1980) Impact of 2,4-D on *Ceuthorhychidius horridus* (Coleoptera: Curculionidae) and their compatibility for integrated control of *Carduus* thistles. *Weed Res.*, **20**: 73-75.

TUEY, D.B. & JAMES, M.O. (1980) Compartmental analysis of 2,4-dichlorophenoxyacetic acid pharmacokinetics in spiny lobster. *Fed. Proc.*, **39**: 525.

VARDIA, H.K. & DURVE, V.S. (1981) The toxicity of 2,4-D to *Cyprinus carpio* var. *communis* in relation to the seasonal variation in the temperature. *Hydrobiologia*, **77**: 155-159.

VARDIA, H.K., RAO, P.S., & DURVE, V.S. (1984) Sensitivity of toad larvae to 2,4-D and endosulfan pesticides. *Arch. Hydrobiol.*, **100**: 395-400.

VERMA, D., CHOUHAN, M.S., & PANDEY, A.K. (1984) Effect of Weedone (a selective 2,4-D herbicide) on neuroendocrine complex of *Puntius ticto* (Ham.). *J. environ. Biol.*, **5**: 249-254.

WATSON, J.R. (1977) Seasonal variation in the biodegradation of 2,4-D in river water. *Water Res.*, **11**: 153-157.

WELP, G. & BRUMMER, G. (1985) [The Fe(III)-reduction test - a simple procedure to determine the effects of environmental chemicals on the microbial activity in soils.] *Z. Pflanzenernaehr. Bodenkd.*, **148**: 10-23 (in German).

WHITEHEAD, C.C. & PETTIGREW, R.J. (1972a) The subacute toxicity of 2,4-dichlorophenoxyacetic acid and 2,4,5-trichlorophenoxyacetic acid to chicks. *Toxicol. appl. Pharmacol.*, **21**: 348-354.

WHITEHEAD, C.C. & PETTIGREW, R.J. (1972b) The effect of 2,4-dichlorophenoxyacetic acid on laying hens. *Br. poult. Sci.*, **13**: 191-195.

WHITNEY, E.W., MONTGOMERY, A.B., MARTIN, E.C., & GANGSTAD, E.O. (1973) The effects of a 2,4-D application on the biota and water quality in Currituck Sound, North Carolina. *Hyacinth Control J.*, **11**: 13-17.

WHO (1984) *Environmental Health Criteria 29: 2,4-dichlorophenoxy-acetic acid (2,4-D)*, Geneva, World Health Organization, 151 pp.

WOODWARD, D.F. (1982) Acute toxicity of mixtures of range management herbicides to cutthroat trout. *J. Range Manage.*, **35**: 539-540.

YOUNG, A.L., THALKEN, C.E., & WARD, W.E. (1975) *Studies of the ecological impact of repetitive aerial applications of herbicides on the ecosystem of Test Area C-52A, Elgin AFB, Florida.* Florida Air Force Armament Laboratory, Elgin Air Force Base, 126 pp. (Report No. AFATL-TR-75-142).

Other titles available in the ENVIRONMENTAL HEALTH CRITERIA series (continued):

44. Mirex
45. Camphechlor
46. Guidelines for the Study of Genetic Effects in Human Populations
47. Summary Report on the Evaluation of Short-term Tests for Carcinogens (Collaborative Study on *In Vitro* Tests)
48. Dimethyl Sulfate
49. Acrylamide
50. Trichloroethylene
51. Guide to Short-term Tests for Detecting Mutagenic and Carcinogenic Chemicals
52. Toluene
53. Asbestos and Other Natural Mineral Fibres
54. Ammonia
55. Ethylene Oxide
56. Propylene Oxide
57. Principles of Toxicokinetic Studies
58. Selenium
59. Principles for Evaluating Health Risks from Chemicals During Infancy and Early Childhood: The Need for a Special Approach
60. Principles and Methods for the Assessment of Neurotoxicity Associated With Exposure to Chemicals
61. Chromium
62. 1,2-Dichloroethane
63. Organophosphorus Insecticides - A General Introduction
64. Carbamate Pesticides - A General Introduction
65. Butanols - Four Isomers
66. Kelevan
67. Tetradifon
68. Hydrazine
69. Magnetic Fields
70. Principles for the Safety Assessment of Food Additives and Contaminants in Food
71. Pentachlorophenol
72. Principles of Studies on Diseases of Suspected Chemical Etiology and Their Prevention
73. Phosphine and Selected Metal Phosphides
74. Diaminotoluenes
75. Toluene Diisocyanates
76. Thiocarbamate Pesticides - A General Introduction
77. Man-made Mineral Fibres
78. Dithiocarbamate Pesticides
79. Dichlorvos
80. Pyrrolizidine Alkaloids
81. Vanadium
82. Cypermethrin
83. DDT and its Derivatives - Environmental Aspects